Laboratory Manual

SIXTH EDITION

ELECTRONIC DEVICES AND CIRCUIT THEORY

ROBERT L. BOYLESTAD
LOUIS NASHELSKY

PRENTICE HALL
Englewood Cliffs, New Jersey   Columbus, Ohio

Editor: Dave Garza/Judith Casillo
Production Editor: Rex Davidson
Cover Designer: Brian Deep
Cover Design Coordinator: Jill E. Bonar
Production Manager: Laura Messerly

This book was set in Times Roman by The Special Projects Group and was printed and bound by The Banta Company. The cover was printed by Phoenix Color Corp.

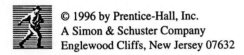 © 1996 by Prentice-Hall, Inc.
A Simon & Schuster Company
Englewood Cliffs, New Jersey 07632

All rights reserved. No part of this book may be
reproduced, in any form or by any means,
without permission in writing from the publisher.

Printed in the United States of America
10 9 8 7 6 5 4 3 2

ISBN: 0-13-375742-0

Prentice-Hall International (UK) Limited, *London*
Prentice-Hall of Australia Pty. Limited, *Sydney*
Prentice-Hall Canada Inc., *Toronto*
Prentice-Hall Hispanoamericana, S.A., *Mexico*
Prentice-Hall of India Private Limited, *New Delhi*
Prentice-Hall of Japan, Inc., *Tokyo*
Simon & Schuster Asia Pte. Ltd., *Singapore*
Editora Prentice-Hall do Brasil, Ltda., *Rio de Janeiro*

Contents

PREFACE		*v*
EQUIPMENT LIST		*vi*
EXPERIMENT 1	OSCILLOSCOPE AND FUNCTION GENERATOR OPERATION	1
EXPERIMENT 2	DIODE CHARACTERISTICS	13
EXPERIMENT 3	SERIES AND PARALLEL DIODE CONFIGURATIONS	23
EXPERIMENT 4	HALF-WAVE AND FULL-WAVE RECTIFICATION	33
EXPERIMENT 5	CLIPPING CIRCUITS	47
EXPERIMENT 6	CLAMPING CIRCUITS	59
EXPERIMENT 7	LIGHT-EMITTING AND ZENER DIODES	73
EXPERIMENT 8	BIPOLAR JUNCTION TRANSISTOR (BJT) CHRACTERISTICS	85
EXPERIMENT 9	FIXED- AND VOLTAGE-DIVIDER BIAS OF A BJT	95
EXPERIMENT 10	EMITTER AND COLLECTOR FEEDBACK BIAS OF BJTs	107
EXPERIMENT 11	DESIGN OF BJT BIAS CIRCUITS	123
EXPERIMENT 12	JFET CHARACTERISTICS	141
EXPERIMENT 13	JFET BIAS CIRCUITS	153
EXPERIMENT 14	DESIGN OF JFET BIAS CIRCUITS	163
EXPERIMENT 15	COMPOUND CONFIGURATIONS	177

EXPERIMENT 16	MEASUREMENT TECHNIQUES	189
EXPERIMENT 17	COMMON-EMITTER TRANSISTOR AMPLIFIER	205
EXPERIMENT 18	COMMON-BASE AND EMITTER-FOLLOWER (COMMON-COLLECTOR) TRANSISTOR AMPLIFIERS	213
EXPERIMENT 19	DESIGN OF COMMON-EMITTER AMPLIFIER	227
EXPERIMENT 20	COMMON-SOURCE TRANSISTOR AMPLIFIER	233
EXPERIMENT 21	MULTISTAGE AMPLIFIER: RC COUPLING	241
EXPERIMENT 22	CMOS CIRCUITS	253
EXPERIMENT 23	DARLINGTON AND CASCODE AMPLIFIER CIRCUITS	259
EXPERIMENT 24	CURRENT SOURCE AND CURRENT MIRROR CIRCUITS	267
EXPERIMENT 25	FREQUENCY RESPONSE OF A COMMON-EMITTER AMPLIFIER	275
EXPERIMENT 26	CLASS-A AND CLASS-B POWER AMPLIFIERS	283
EXPERIMENT 27	DIFFERENTIAL AMPLIFIER CIRCUITS	291
EXPERIMENT 28	LINEAR OP-AMP CIRCUITS	303
EXPERIMENT 29	ACTIVE FILTER CIRCUITS	311
EXPERIMENT 30	COMPARATOR CIRCUIT OPERATION	319
EXPERIMENT 31	OSCILLATOR CIRCUITS	327
EXPERIMENT 32	VOLTAGE REGULATION-POWER SUPPLIES	337

Preface

For this edition our effort was directed toward ensuring that the data obtained was meaningful and the instructions clear. Each experiment has now been class tested for the past three years and should be in good form. Although some experiments reflect suggestions submitted by reviewers and warrants users, the heading for each remains the same.

The first half of the manual is devoted primarily to dc analysis of electronic circuits with the second half devoted to ac operation of electronic circuits.

Graphs are provided for all the experiments for plotting data or recording waveforms. In addition, space is set aside for calculations and to answer questions. In general, when the experiment is completed it can be removed from the laboratory manual at the perforations and submitted for grading. There should seldom be need for additional pages to report on the experiment.

An equipment list is provided to help determine the availability of equipment to run all the laboratory experiments. A sincere effort was made to limit the parts list and to keep the power levels as low as possible. Lists showing parts needed for Experiments 1-15, for Experiments 16-32, or for all experiments are provided.

The authors wish to extend their sincerest appreciation to Professors Bill Boettcher, Jake Froese, Doug Fuller, Lee Rosenthal, and Gerald Terrebrood for their excellent reviews and very helpful suggestions. In addition, we thank Judy Casillo of Prentice-Hall, Inc. for her excellent work in preparing this manual.

Robert Boylestad
Louis Nashelsky

Equipment List

EXPERIMENTS 1-15

INSTRUMENTS

Oscilloscope (dual trace preferable)
Digital Multimeter (DMM)
DC Power Supply
Signal Generator
Frequency Counter

RESISTORS*

(1) 100 Ω	(1) 2 kΩ	(2) 15 kΩ
(1) 220 Ω	(1) 2.2 kΩ	(1) 33 kΩ
(1) 300 Ω	(1) 2.4 kΩ	(1) 100 kΩ
(1) 330 Ω	(1) 2.7 kΩ	(1) 330 kΩ
(1) 470 Ω	(1) 3 kΩ	(1) 390 kΩ
(1) 680 Ω	(1) 3.3 kΩ	(1) 1 MΩ
(2) 1 kΩ	(1) 3.9 kΩ	(1) 10 MΩ
(1) 1.2 kΩ	(1) 4.7 kΩ	
(1) 1.5 kΩ	(1) 6.8 kΩ	
(1) 1.8 kΩ	(1) 10 kΩ	

(1) 1 kΩ potentiometer
(1) 5 kΩ potentiometer
(1) 1 MΩ potentiometer

---

*all resistors 1/2 W, unless otherwise indicated

CAPACITORS

(1) 0.1 µF
(1) 1 µF

DIODES

(4) Silicon
(1) Germanium
(1) LED
(1) Zener (10 V)

TRANSISTORS

(2) BJT 2N3904 (or equivalent)
(1) BJT 2N4401 (or equivalent)
(1) JFET 2N4416 (or equivalent)
(1) BJT without terminal identification

MISCELLANEOUS

(1) 12.6 V Center-tapped transformer with fused line cord
(1) Heat gun (if available)
(1) Curve tracer (if available)

EXPERIMENTS 16-32

INSTRUMENTS

Oscilloscope (dual trace preferable)
Digital Multimeter (DMM)
DC Power Supply
Signal Generator
Frequency Counter

RESISTORS*

(1) 20 Ω	(2) 1 kΩ	(1) 4.3 kΩ	(1) 39 kΩ
(1) 51 Ω, 1 W	(2) 1.2 kΩ	(1) 4.7 kΩ	(1) 51 kΩ
(1) 82 Ω	(1) 1.8 kΩ	(1) 5.1 kΩ	(3) 100 kΩ
(1) 120 Ω, 0.5 W	(1) 2 kΩ	(1) 5.6 kΩ	(1) 220 kΩ
(1) 150 Ω	(2) 2.2 kΩ	(1) 6.8 kΩ	(2) 1 MΩ
(1) 180 Ω	(2) 2.4 kΩ	(1) 7.5 kΩ	
(2) 390 Ω, 0.5 W	(2) 3 kΩ	(5) 10 kΩ	
(1) 510 Ω	(1) 3.3 kΩ	(2) 20 kΩ	
(1) 1 kΩ, 0.5 W	(1) 3.9 kΩ	(1) 33 kΩ	

(1) 50 kΩ potentiometer
(1) 500 kΩ potentiometer

---

*all resistors 1/2 W, unless otherwise indicated

CAPACITORS

(3) 0.001 µF
(3) 0.01 µF
(1) 0.1 µF
(1) 1 µF
(4) 10 µF
(3) 15 µF
(2) 20 µF
(2) 100 µF

DIODES

(2) Silicon
(1) LED (20 mA)

TRANSISTORS

(3) BJT 2N3904 (or equivalent)
(3) JFET 2N3823 (or equivalent)
(1) TIP 120
(1) 2N4300 npn medium power
(1) 2N5333 pnp medium power

ICs

(1) 74HC02 or 14002 CMOS gate
(1) 74HC04 or 14004 CMOS inverter
(1) 7414 Schmitt-trigger hex inverter
(1) 301 op-amp
(1) 339 comparator
(1) 741 op-amp

ALL EXPERIMENTS

INSTRUMENTS

Oscilloscope (dual trace preferable)
Digital Multimeter (DMM)
DC Power Supply
Signal Generator
Frequency Counter

RESISTORS*

(1) 20 Ω	(2) 1 kΩ	(1) 5.6 kΩ
(1) 51 Ω, 1 W	(2) 1 kΩ, 0.5 W	(1) 6.8 kΩ
(1) 82 Ω	(2) 1.2 kΩ	(1) 7.5 kΩ

---

*all resistors 1/2 W, unless otherwise indicated

(1) 100 Ω	(1) 1.8 kΩ	(5) 10 kΩ
(1) 120 Ω, 0.5 W	(1) 2 kΩ	(2) 15 kΩ
(1) 150 Ω	(2) 2.2 kΩ	(2) 20 kΩ
(1) 180 Ω	(2) 2.4 kΩ	(1) 33 kΩ
(1) 220 Ω	(1) 2.7 kΩ	(1) 39 kΩ
(1) 300 Ω	(2) 3 kΩ	(1) 51 kΩ
(1) 330 kΩ	(1) 3.3 kΩ	(3) 100 kΩ
(1) 390 Ω, 0.5 W	(1) 3.9 kΩ	(1) 220 kΩ
(1) 470 Ω	(1) 4.3 kΩ	(1) 330 kΩ
(2) 510 Ω	(1) 4.7 kΩ	(1) 390 kΩ
(1) 680 Ω	(5) 5.1 kΩ	(2) 1 MΩ
		(1) 10 MΩ

(1) 1 kΩ potentiometer
(1) 5 kΩ potentiometer
(1) 50 kΩ potentiometer
(1) 500 kΩ potentiometer
(1) 1 MΩ potentiometer

CAPACITORS

(3) 0.001 µF
(3) 0.01 µF
(1) 0.1 µF
(1) 1 µF
(4) 10 µF
(3) 15 µF
(2) 20 µF
(2) 100 µF

DIODES

(4) Silicon
(1) Germanium
(1) LED (20 mA)
(1) Zener (10 V)

TRANSISTORS

(3) BJT 2N3904 (or equivalent)
(3) JFET 2N3823 (or equivalent)
(1) BJT 2N4401 (or equivalent)
(1) JFET 2N4416 (or equivalent)
(1) TIP 120
(1) 2n4300 npn medium power
(1) 2N5333 pnp medium power
(1) BJT without terminal identification

x Equipment List

ICs

(1) 74HC02 or 14002 CMOS gate
(1) 74HC04 or 14004 CMOS inverter
(1) 7414 Schmitt-trigger hex inverter
(1) 301 Op-amp
(1) 339 comparator
(1) 741 Op-amp

MISCELLANEOUS

(1) 12.6 V Center-tapped transformer with fused line cord
(1) Heat gun (if available)
(1) Curve tracer (if available)

Name _____
Date _____
Instructor _____

EXPERIMENT 2

Diode Characteristics

OBJECTIVE

To become familiar with the characteristics of a silicon and germanium diode.

EQUIPMENT REQUIRED

Instruments

DMM

Components

Resistors

(1) 1-kΩ
(1) 1-MΩ

Diodes

(1) Silicon
(1) Germanium

Supplies

DC power supply

Miscellaneous

Demonstration: 1 heat gun

Exp. 2 / Diode Characteristics

EQUIPMENT ISSUED

Item	Laboratory serial no.
DMM	
DC power supply	

RÉSUMÉ OF THEORY

Most modern-day digital multimeters can be used to determine the condition of a diode. They have a scale denoted by a diode symbol that will indicate the condition of a diode in the forward and reverse-bias regions. If connected to establish a forward-bias condition the meter will display the forward voltage across the diode at a current level typically in the neighborhood of 2 mA. If connected to establish a reverse-bias condition an "OL" should appear on the display to support the open-circuit approximation frequently applied to this region. If the meter does not have the diode-checking capability the condition of the diode can also be checked by obtaining some measure of the resistance level in the forward and reverse-bias regions. Both techniques for checking a diode will be introduced in the first part of the experiment.

The characteristics of a silicon or germanium diode have the general shape shown in Fig. 2.1. Note the change in scale for both the vertical and horizontal axes. In the reverse-biased region the reverse saturation currents are fairly constant from 0 V to the Zener potential. In the forward-bias region the current increases quite rapidly with increasing diode voltage. Note that the curve is rising almost vertically at a forward-biased voltage of less than 1 V. The forward-biased diode current will be limited solely by the network in which the diode is connected or by the maximum current or power rating of the diode.

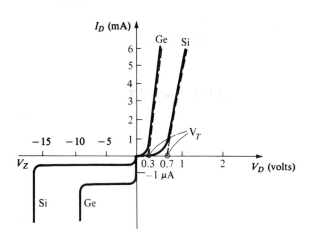

The "firing potential" or threshold voltage is determined by extending a straight line (dashed lines of Fig. 2.1) tangent to the curves until it hits the horizontal axis. The intersection with the V_D axis will determine the threshold voltage V_T.

Figure 2.1 Silicon and germanium diode characteristics.

The *DC* or *Static resistance* of a diode at any point on the characteristics is determined by the ratio of the diode voltage at that point, divided by the diode current. That is,

$$R_{DC} = \frac{V_D}{I_D} \quad \text{ohms} \tag{2.1}$$

The *AC resistance* at a particular diode current or voltage can be determined using a tangent line drawn as shown in Fig. 2.1. The resulting voltage (ΔV) and current (ΔI) deviations can then be determined and the following equation applied.

$$r_d = \frac{\Delta V}{\Delta I} \quad \text{ohms} \tag{2.2}$$

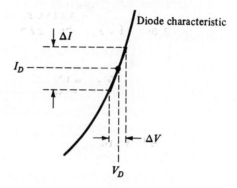

Figure 2-2

It can be shown through differential calculus that the AC resistance of a diode in the vertical-rise section of the characteristics is given by

$$r_d = \frac{26 \text{ mV}}{I_D} \quad \text{ohms} \tag{2.3}$$

For levels of current at and below the knee of the curve the AC resistance of a silicon diode is better approximated by

$$r_d = 2\left(\frac{26 \text{ mV}}{I_D}\right) \quad \text{ohms} \tag{2.4}$$

PROCEDURE

Part 1. Diode Test

Diode Testing Scale

The diode-testing scale of the DMM can be used to determine the condition of a diode. With one polarity, the DMM should provide the "firing potential" of the diode, while the reverse connection should result in an "OL" response to support the open-circuit approximation.

Using the connection in Fig. 2.2, constant-current source of about 2 mA internal to the meter will forward bias the junction, and a voltage in the neighborhood of 0.7 V (700 mV) should be obtained for silicon and 0.3 V (300 mV) for germanium. If the leads are reversed, an OL indication should be obtained.

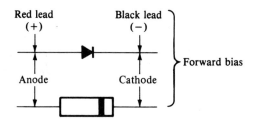

Figure 2.3 Diode testing.

If a low reading (less than 1 V) is obtained in both directions, the junction is shorted internally. If an OL indication is obtained in both directions, the junction is open.

Perform the tests of Table 2.1 for the silicon and germanium diodes.

TABLE 2.1

Test	Si	Ge
Forward		
Reverse		

Based on the results of Table 2.1, are both diodes in good condition?

Resistance Scales

As indicated in the Résumé of Theory the condition of a diode can also be checked using the resistance scales of a VOM or digital meter. Using the appropriate scales of the VOM or DMM, determine the resistance levels of the forward- and reverse-bias regions of the Si and Ge diodes. Enter the results in Table 2.2.

TABLE 2.2

Test	Si	Ge
Forward		
Reverse		

Although the firing potential is not revealed using the resistance scales, a "good" diode will result in a lower resistance level in the forward bias state and a much higher resistance level when reverse-biased.

Based on the results of Table 2.2, are both diodes in good condition?

Part 2. Forward-bias Diode Characteristics

In this part of the experiment we will obtain sufficient data to plot the forward-bias characteristics of the silicon and germanium diodes on Fig. 2.5.

    a. Construct the network of Fig. 2.4 with the supply (E) set at 0 V. Record the measured value of the resistor.

Procedure

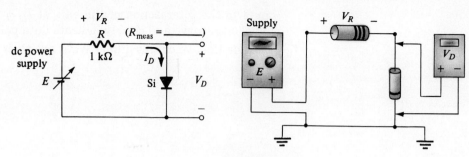

Figure 2-4

b. Increase the supply voltage until V_R (not E) reads 0.1 V. Then measure V_D and insert in Table 2.3. Calculate I_D using the equation shown in Table 2.3.

TABLE 2.3
V_D versus I_D for the silicon diode

V_R (V)	0.1	0.2	0.3	0.4	0.5	0.6	0.7	0.8
V_D (V)								
$I_D = \dfrac{V_R}{R_{meas}}$ (mA)								

V_R (V)	0.9	1	2	3	4	5	6	7	8	9	10
V_D (V)											
$I_D = \dfrac{V_R}{R_{meas}}$ (mA)											

c. Repeat step b for the remaining settings of V_R.
d. Replace the silicon diode by a germanium diode and complete Table 2.4.

TABLE 2.4
V_D versus I_D for the germanium diode

V_R (V)	0.1	0.2	0.3	0.4	0.5	0.6	0.7	0.8
V_D (V)								
$I_D = \dfrac{V_R}{R_{meas}}$ (mA)								

V_R (V)	0.9	1	2	3	4	5	6	7	8	9	10
V_D (V)											
$I_D = \dfrac{V_R}{R_{meas}}$ (mA)											

e. On Fig. 2.5, plot I_D versus V_D for the silicon and germanium diodes. Finish off the curves by extending the lower region of the curve to the intersection of the axis at $I_D = 0$ mA and $V_D = 0$ V. Label each curve and clearly indicate data points. Be neat!

f. How do the two curves differ? What are their similarities?

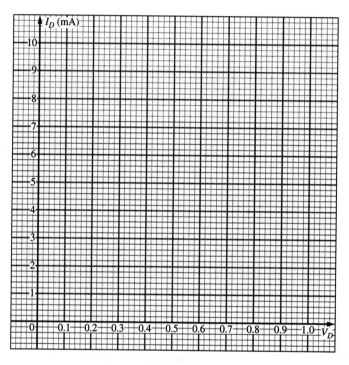

Figure 2-5

Part 3. Reverse Bias

a. In Fig. 2.6 a reverse-bias condition has been established. Since the reverse saturation current will be relatively small, a large resistance of 1 MΩ is required if the voltage across R is to be of measurable dimensions. Construct the circuit of Fig. 2.6 and record the measured value of R on the diagram.

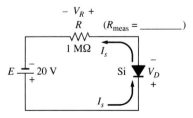

Figure 2-6

b. Measure the voltage V_R. Calculate the reverse saturation current from $I_s = V_R/(R_{meas}||R_m)$. The internal resistance (R_m) of the DMM is included because of the large magnitude of the resistance R. Your instructor will provide the internal resistance of the DMM for your calculations. If unavailable, use a typical value of 10 MΩ.

$R_m =$ _____

(measured) $V_R =$ _____

(calculated) $I_s =$ _____

c. Repeat step 3(b) for the germanium diode.

(measured) $V_R =$ _____

(calculated) $I_s =$ _____

d. How do the resulting levels of I_s for silicon and germanium compare?

e. Determine the DC resistance levels for the silicon and germanium diodes using the equation

$$R_{DC} = \frac{V_D}{I_D} = \frac{V_D}{I_s} = \frac{E - V_R}{I_s}$$

(calculated) R_{DC} (Si) = _____

(calculated) R_{DC} (Ge) = _____

Are the resistance levels sufficiently high to be considered open-circuit equivalents if appearing in series with resistors in the low kilohm range?

Part 4. DC Resistance

a. Using the Si curve of Fig. 2.5 determine the diode voltage at diode current levels indicated in Table 2.5. Then determine the DC resistance at each current level. Show all calculations.

TABLE 2.5

I_D(mA)	V_D	R_{DC}
0.2		
1		
5		
10		

b. Repeat Part 4(a) for germanium and complete Table 2.6 (Table 2.6 is the same as Table 2.5).

TABLE 2.6

I_D(mA)	V_D	R_{DC}
0.2		
1		
5		
10		

c. Are there any trends in DC resistance (for Si and Ge) as the diode current increases and we move up the vertical-rise section of the characteristics?

Part 5. AC Resistance

a. Using the equation $r_d = \Delta V/\Delta I$ [Eq. (2.2)], determine the AC resistance of the silicon diode at I_D = 9 mA using the curve of Fig. 2.5. Show all work.

(calculated) r_d = _____

Procedure

b. Determine the AC resistance at $I_D = 9$ mA using the equation $r_d = 26$ mV/I_D (mA) for the silicon diode. Show all work.

(calculated) $r_d =$ _____

How do the results of part (a) and (b) compare?

c. Repeat step 5(**a**) for $I_D = 2$ mA for the silicon diode.

(calculated) $r_d =$ _____

d. Repeat step 5(**b**) for $I_D = 2$ mA for the silicon diode. Use Eq. 2.4.

(calculated) $r_d =$ _____

How do the results of Part 5(**c**) and 5(**d**) compare?

Part 6. Firing Potential

Graphically determine the firing potential (threshold voltage) of each diode from its characteristics as defined in the Résumé of Theory. Show the straight-line approximations on Fig. 2.5.

V_T (silicon) = _____

V_T (germanium) = _____

Part 7. Temperature Effects (Demonstration)

Reconstruct the circuit of Fig. 2.4 using the silicon diode. Establish a current of about 1 mA by setting V_R to 1 V.

a. Place the voltmeter section on the DMM across the diode and note the reading as the instructor heats the diode with the heat gun. Record the effect on V_D of heating the diode.

b. Let the diode cool down and then place the voltmeter section across the resistor R. Note the effect on V_R of heating the diode. Since $I_D = V_R/R$ what effect on the diode current of the network results from heating the diode?

c. Since $R_{\text{diode}} = V_D/I_D$ what is the effect of increasing temperature on the resistance of the diode?

d. Does a semiconductor diode have a positive or negative temperature coefficient? Explain.

Questions

1. Compare the characteristics of silicon and germanium in the forward- and reverse-bias regions. In particular, which diode is closer to the short-circuit approximation in the forward-bias region and which is closer to the open-circuit approximation in the reverse-bias region? How are they similar and what are their most noticeable differences?

2. Research the effect of heat on the terminal resistance of semiconductor materials and briefly review why the terminal resistance will decrease with the application of heat.

	Name _____
	Date _____
	Instructor _____

EXPERIMENT 3

Series and Parallel Diode Configurations

OBJECTIVE

To develop the ability to analyze networks with diodes in a series or parallel configuration.

EQUIPMENT REQUIRED

Instruments

DMM

Components

Resistors

(1) 1-kΩ
(2) 2.2-kΩ

Diodes

(2) Silicon
(1) Germanium

Supplies

DC power supply

EQUIPMENT ISSUED

Item	Laboratory serial no.
DC power supply	
DMM	

RÉSUMÉ OF THEORY

The analysis of circuits with diodes and a DC input requires that the state of the diodes first be determined. For silicon diodes (with a transition voltage or "firing potential" of 0.7 V), the voltage across the diode must be at least 0.7 V with the polarity appearing in Fig. 3.1a for the diode to be in the "on" state. Once the voltage across the diode reaches 0.7 V the diode will turn "on" and have the equivalent of Fig. 3.1b. For $V_D < 0.7$ V or for voltages with the opposite polarity of Fig. 3.1a, the diode can be approximated as an open circuit. For germanium diodes, simply replace the transition voltage by the germanium value of 0.3 V.

In most networks where the applied DC voltage exceeds the transition voltage of the diodes, the state of the diode can usually be determined simply by mentally replacing the diode by a resistor and determining the direction of current through the resistor. If the direction matches the arrowhead of the diode symbol, the diode is in the "on" state, and if the opposite, it is in the "off" state. Once the state is determined, simply replace the diode by the transition voltage or open circuit and analyze the rest of the network.

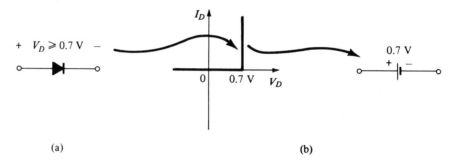

(a)   (b)

Figure 3-1 Forward-biased silicon diode.

Be continually alert to the location of the output voltage $V_o = V_R = I_R R$. This is particularly helpful in situations where a diode is in an open-circuit condition and the current is zero. For $I_R = 0$, $V_o = V_R = I_R R = 0(R) = 0$ V. In addition, recall that an open circuit can have a voltage across it, but the current is zero. Further, a short circuit has a zero-volt drop across it, but the current is limited only by the external network or limitations of the diode.

The analysis of logic gates requires that one make an assumption about the state of the diodes, determine the various voltage levels, and then determine whether the results violate any basic laws, such as the fact that a point in a network (such as V_o) can have only one voltage level. It is usually helpful to keep in mind that there must be a forward-bias voltage across a diode equal to the transition voltage to turn it "on." Once V_o is determined and no laws are violated with the diodes in their assumed state, a solution to the configuration can be assumed.

PROCEDURE

PART 1. Threshold Voltage V_T

For both the silicon and the germanium diode, determine the threshold using the diode-checking capability of the DMM or a curve tracor. For this experiment the "firing voltages" obtained will establish the equivalent characteristics for each diode appearing in Fig. 3.2. Record the value of V_T

Procedure

obtained for each diode in Fig. 3.2. If the diode checking capability or curve tracer is unavailable, assume $V_T = 0.7$ V for silicon and $V_T = 0.3$ V for germanium.

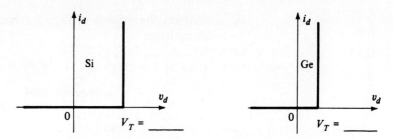

Figure 3-2 Firing voltage for silicon and germanium.

Part 2. Series configuration

a. Construct the circuit of Fig 3.3. Record the measured value of R.

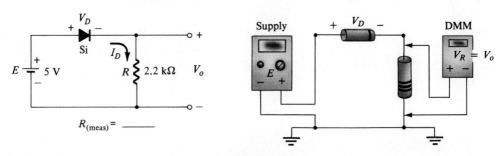

Figure 3-3

b. Using the results of Part 1 and the measured resistance for R, calculate the theoretical values of V_o and I_D. Insert the level of V_T for V_D.

$V_D =$ _____

(calculated) $V_o =$ _____

(calculated) $I_D =$ _____

c. Measure the voltages V_D and V_o, using the DMM. Calculate the current I_D from measured values. Compare with the results of part 2(**b**).

(measured) V_D = _____

(measured) V_o = _____

(from measured) $I_D = \dfrac{V_o}{R}$ = _____

d. Construct the circuit of Fig. 3.4 Record the measured values for each resistor.

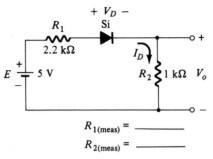

$R_{1(meas)}$ = _____

$R_{2(meas)}$ = _____

Figure 3-4

e. Using the results of Part 1 and the measured resistance values for R_1 and R_2, calculate the theoretical values of V_o and I_D. Insert the level of V_T for V_D.

V_D = _____

(calculated) V_o = _____

(calculated) I_D = _____

f. Measure the voltages V_D and V_o, using the DMM. Calculate the current I_D from measured values. Compare with the results of step 2(**e**).

(measured) V_D = _____

(measured) V_o = _____

(from measured) $I_D = \dfrac{V_o}{R_2}$ = _____

g. Reverse the silicon diode in Fig. 3.4 and calculate the theoretical values of V_D, V_o and I_D.

Procedure

$V_D =$ _____
(calculated) $V_o =$ _____
(calculated) $I_D =$ _____

h. Measure V_D and V_o, for the conditions of step **h**. Calculate the current I_D from measured values. Compare with the results of Part 2(**g**).

(measured) $V_D =$ _____
(measured) $V_o =$ _____
(from measured) $I_D = \dfrac{V_o}{R_2} =$ _____

i. Construct the network of Fig. 3.5. Record the measured value of R.

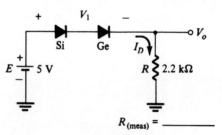

Figure 3-5

j. Using the results of Part 1, calculate the theoretical values of V_1 (across both diodes), V_o, and I_D.

(calculated) $V_1 =$ _____
(calculated) $V_o =$ _____
(calculated) $I_D =$ _____

k. Measure V_1 and V_o, and compare against the results of step **j**. Calculate the current I_D from measured values and compare to the level of Part 2(**j**).

(measured) $V_1 =$ _____

(measured) $V_o =$ _____

(from measured) $I_D = \dfrac{V_o}{R} =$ _____

Part 3. Parallel configuration

a. Construct the network of Fig. 3.6. Record the measured value of R.

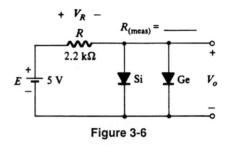

Figure 3-6

b. Using the results of Part 1, calculate the theoretical values of V_o and V_R.

(calculated)) $V_o =$ _____

(calculated)) $V_R =$ _____

c. Measure V_o and V_R and compare with the results of step 3(b).

(measured) $V_o =$ _____

(measured) $V_R =$ _____

d. Construct the network of Fig. 3.7. Record the measured value of each resistor.

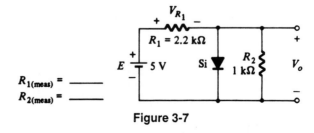

Figure 3-7

Procedure

e. Using the results of step a of Part 1, calculate the theoretical values of V_o, V_{R_1}, and I_D.

(calculated) $V_o =$ _____
(calculated) $V_{R_1} =$ _____
(calculated) $I_D =$ _____

f. Measure V_o and V_{R_1}. Using the measured values of V_o and V_{R_1} calculate I_{R_2} and I_{R_1} and determine I_D. Compare to the results of Part 3(e).

(measured) $V_o =$ _____
(measured) $V_{R_1} =$ _____
(from measured) $I_D =$ _____

g. Construct the network of Fig. 3.8. Record the measured value of the resistor.

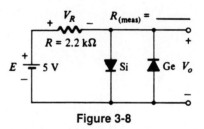

Figure 3-8

h. Using the results of Part 1, calculate the theoretical values of V_o and V_R.

(calculated) $V_o =$ _____
(calculated) $V_R =$ _____

i. Measure V_o and V_R and compare with the results of step 3(h).

(measured) V_o = _____
(measured) V_R = _____

Part 4. Positive Logic AND Gate

a. Construct the network of Fig 3.9. Record the measured value of the resistor.

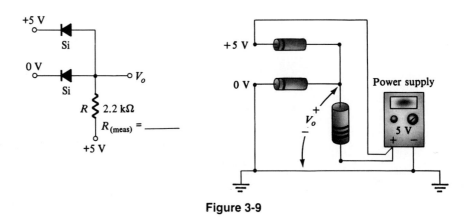

Figure 3-9

b. Using the V_T of Part 1, calculate the theoretical value of V_o.

(calculated) V_o = _____

c. Measure V_o and compare to step 4(**b**).

(measured) V_o = _____

d. Apply 5 V to each input terminal of Fig. 3.9 and calculate the theoretical value of V_o.

(calculated) V_o = _____

e. Measure V_o and compare to the results of step 4(**d**).

(measured) V_o = _____

Procedure

f. Set both inputs to zero in Fig. 3.9 (by connecting both inputs to circuit ground) and calculate the theoretical value of V_o.

(calculated) $V_o =$ _____

g. Measure V_o and compare to the results of step 4(**f**).

(measured) $V_o =$ _____

Part 5. Bridge configuration

a. Construct the network of Fig. 3.10. Record the measured value of each resistor.

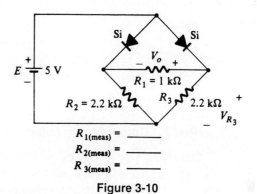

Figure 3-10

b. Using the V_T of Part 1, calculate the theoretical value of V_o and V_{R_3}.

(calculated) $V_o =$ _____
(calculated) $V_{R_3} =$ _____

c. Measure V_o and V_{R_3} and compare to the results of step 5(**b**). Use a low voltage scale when measuring V_o.

(measured) $V_o =$ _____
(measured) $V_{R_3} =$ _____

Part 6. Practical Exercise

    a. If the diode in the top right branch of Fig. 3.10 were damaged, creating an internal open-circuit, calculate the resulting levels of V_o and V_{R_3}.

(calculated) $V_o =$ _____

(calculated) $V_{R_3} =$ _____

    b. Remove the top right diode from Fig. 3.10 and measure V_o and V_{R_3}. Compare the results with those predicted in Part 6(a).

(measured) $V_o =$ _____

(measured) $V_{R_3} =$ _____

Part 7. Computer Exercise

Analyze the network of Fig. 3.4 using PSpice, MicroCap II, or other appropriate software package.

    For PSpice the input file is the following:

```
DIODE CIRCUIT OF FIGURE 3.4
VE     1    0    5V
R1     1    2    2.2K
D1     2    3    DI
R2     3    0    1K
.MODEL DI D(IS=2E-15)
.DC VE   6V   6V   1V
.PRINT DC V(R2), I (D1)
.OPTIONS NOPAGE
.END
```

Compare the results with those obtained in Part 2(**f**).

Computer $V_o =$ _____   [Part 2(**f**)] $V_o =$ _____

Computer $I_D =$ _____   [Part 2(**f**)] $I_D =$ _____

Name _____
Date _____
Instructor _____

EXPERIMENT 4

Half-Wave and Full-Wave Rectification

OBJECTIVE

To become familiar with half-wave and full-wave rectification.

EQUIPMENT REQUIRED

Instruments

Oscilloscope
DMM

Components

Resistors

(2) 2.2-kΩ
(1) 3.3-kΩ

Diodes

(4) Silicon

Supplies

Function generator

Miscellaneous

12.6-V Center-tapped transformer with fused line cord

Exp. 4 / Half-Wave and Full-Wave Rectification

EQUIPMENT ISSUED

Item	Laboratory serial no.
Oscilloscope	
DMM	
Function generator	

RÉSUMÉ OF THEORY

The primary function of half-wave and full-wave rectification systems is to establish a DC level from a sinusoidal input signal that has zero average (DC) level.

The half-wave signal of Fig. 4.1 normally established by a network with a single diode has an average or equivalent DC level equal to 31.8% of the peak value V_m.

That is,

$$V_{dc} = 0.318 V_{peak} \quad \text{(4.1)}$$
half-wave

The full-wave rectified signal of Fig. 4.2 has twice the average or DC level of the half-wave signal, or 63.6% of the peak value V_m.

That is,

$$V_{dc} = 0.636 V_{peak} \quad \text{(4.2)}$$
full-wave

For large sinusoidal inputs ($V_m \gg V_T$) the forward-biased transition voltage of a diode can be ignored. However, for situations when the peak value of the sinusoidal signal is not that much greater than V_T, V_T can have a noticeable effect on V_{DC}.

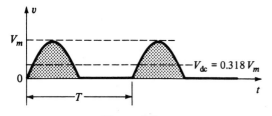

Figure 4-1

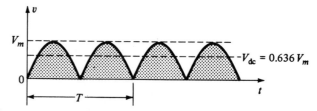

Figure 4-2 Full-wave rectified signal.

In rectification systems the peak inverse voltage (PIV) or Zener breakdown voltage parameter must be considered carefully. For typical single-diode half-wave rectification systems, the required PIV level is equal to the peak value of the applied sinusoidal signal. For the four-diode full-wave bridge rectification system, the required PIV level is again the peak value, but for a two-diode center-tapped configuration, it is twice the peak value of the applied signal. The PIV voltage is the maximum reverse-bias voltage that a diode can handle before entering the Zener breakdown region.

Procedure

PROCEDURE

Part 1. Threshold Voltage

Chose one of the silicon diodes and determine the threshold voltage, V_T, using the diode-checking capability of the DMM or a curve tracer.

$V_T =$ _____

Part 2. Half-Wave Rectification

    a. Construct the circuit of Fig. 4.3 using the chosen diode of Part 1. Record the measured value of the resistance. Set the function generator to a 1000-Hz 8-V p-p sinusoidal voltage using the oscilloscope.

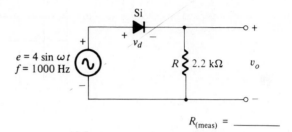

$R_{(meas)} =$ _____

Figure 4.3 Half-wave rectifier.

    b. The sinusoidal input (e) of Fig. 4.3 has been plotted on the screen of Fig. 4.4. Determine the chosen vertical and horizontal sensitivities. Note that that the horizontal axis is the 0 V line.

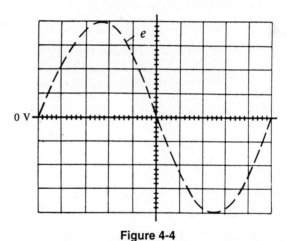

Figure 4-4

Vertical sensitivity = _____

Horizontal sensitivity = _____

    c. Using the threshold voltage of Part 1 determine the the theoretical output voltage v_o for Fig. 4.3 and sketch the waveform on Fig. 4.4

for one full cycle using the same sensitivities employed in Part 2(**b**). Indicate the maximum and minimum values on the output waveform.

d. Using the oscilloscope with the AC-GND-DC coupling switch in the DC position obtain the voltage v_o and sketch the waveform on Fig. 4.5. Before viewing v_o be sure to set the $v_o = 0$ V line using the GND position of the coupling switch. Use the same sensitivities as in Part 2(**b**).

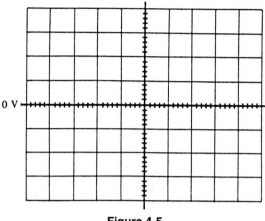

Figure 4-5

How do the results of Parts 2(**c**) and 2(**d**) compare?

e. Calculate the DC level of the half-wave rectified signal of step 2(**d**). Assume the positive pulse of the waveform encompasses one-half the period of the input waveform when using Eq. 4.1.

(calculated) V_{DC} = _____

f. Measure the DC level of v_o using the DC scale of the DMM and find the percent difference between the measured value and the calculated value of Part 2(**e**) using the following equation:

$$\% \text{ Difference} = \left| \frac{V_{DC \text{ (calc)}} - V_{DC \text{ (meas)}}}{V_{DC \text{ (calc)}}} \right| \times 100\%$$

(measured) V_{DC} = _____
(% Difference) = _____

Procedure

g. Switch the AC-GND-DC coupling switch to the AC position. What is the effect on the output signal v_o? Does it appear that the area under the curve above the zero axis equals the area under the curve below the zero axis? Discuss the effect of the AC position on waveforms that have an average value over one full cycle.

h. Reverse the diode of Fig. 4.3 and sketch the output waveform obtained using the oscilloscope on Fig. 4.6. Be sure the coupling switch is in the DC position and the $v_o = 0$ V line is preset using the GND position. Include the maximum and minimum voltage levels on the plot as determined using the chosen vertical sensitivity.

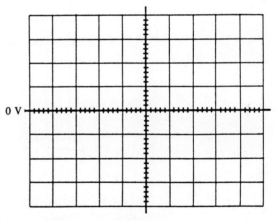

Figure 4-6

i. Calculate and measure the DC level of the resulting waveform of Fig. 4.6. Insert the proper sign for the polarity of V_{DC} as defined by Fig. 4.3. Assume the positive pulse of the waveform encompasses one-half the period of the input waveform when using Eq. 4.1.

(calculated) V_{DC} = _____

(measured) V_{DC} = _____

Part 3. Half-Wave Rectification (continued)

a. Construct the network of Fig. 4.7. Record the measured value of the resistor R.

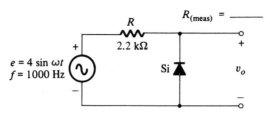

Figure 4-7

b. Using the threshold voltage of Part 1 determine the theoretical output voltage v_o for Fig. 4.7 and sketch the waveform on Fig. 4.8 for one full cycle using the same sensitivities employed in Part 2(**b**). Indicate the maximum and minimum values on the output waveform.

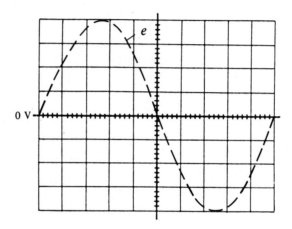

Figure 4-8

c. Using the oscilloscope with the coupling switch in the DC position obtain the voltage v_o and sketch the waveform on Fig. 4.9. Before viewing v_o be sure to set the $v_o = 0$ V line using the GND position of the coupling switch. Use the same sensitivities as in Part 3(**b**).

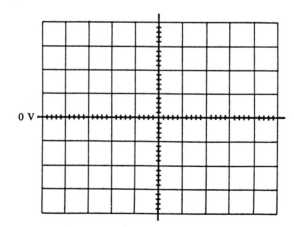

Figure 4-9

How do the results of Parts 3(b) and 3(c) compare?

    d. What is the most noticeable difference between the waveform of Fig. 4.9 and that obtained in Part 2(h)? Why did the difference occur?

    e. Calculate the DC level of the waveform of Fig. 4.9 using the following equation:

$$V_{DC} = \frac{\text{Total Area}}{2\pi} \cong \frac{2V_m - (V_T)\pi}{2\pi} = 0.318V_m - V_T/2$$

(calculated) $V_{DC} =$ _____

    f. Measure the output DC voltage with the DC scale of the DMM and calculate the percent difference using the same equation appearing in Part 2(f).

(measured) $V_{DC} =$ _____
(% Difference) = _____

Part 4. Half-Wave Rectification (continued)

    a. Construct the network of Fig. 4.10. Record the measured value of each resistor.

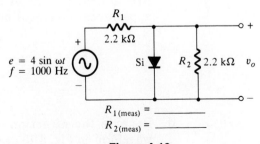

Figure 4-10

b. Using the measured resistor values and V_T from Part 1, forecast the appearance of the output waveform v_o and sketch the result on Fig. 4.11. Use the same sensitivities employed in Part 2(**b**) and insert the maximum and minimum values of the waveform.

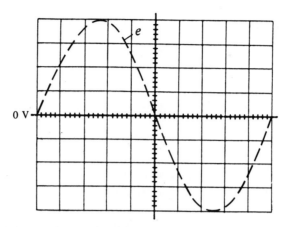

Figure 4-11

c. Using the oscilloscope with the coupling switch in the DC position obtain the waveform for v_o and record on Fig. 4.12. Again, be sure to preset the $v_o = 0$ V line using the GND position of the coupling switch before viewing the waveform. Using the chosen sensitivities determine the maximum and minimum values and place on the sketch of Fig. 4.12.

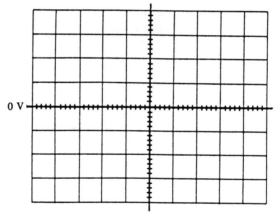

Figure 4-12

Are the waveforms of Figs. 4.11 and 4.12 relatively close in appearance and magnitude?

d. Reverse the direction of the diode and record the resulting waveform on Fig. 4.13 as obtained using the oscilloscope.

Procedure

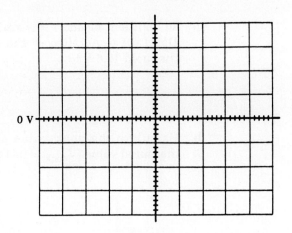

Figure 4-13

Compare the results of Figs. 4.12 and 4.13. What are the major differences and why?

Part 5. Full-Wave Rectification (Bridge configuration)

a. Construct the full-wave bridge rectifier of Fig. 4.14. Be sure that the diodes are inserted correctly and that the grounding is as shown. If unsure, ask your instructor to check your setup. Record the measured value of the resistor R.

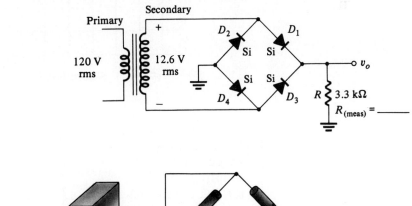

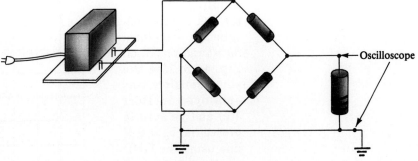

Figure 4-14

Exp. 4 / Half-Wave and Full-Wave Rectification

In addition, measure the rms voltage at the secondary using the DMM set to AC. Record the rms value below. Does it differ from the rated 12.6 V?

(measured) V_{rms} = _____

b. Calculate the peak value of the secondary voltage using the measured value ($V_{peak} = 1.414\, V_{rms}$).

(calculated) V_{peak} = _____

c. Using the V_T of Part 1 for each diode sketch the expected output waveform v_o on Fig. 4.15. Choose a vertical and horizontal sensitivity commensurate with the secondary voltage. Consult your oscilloscope to obtain a list of possibilities. Record your choice for each below.

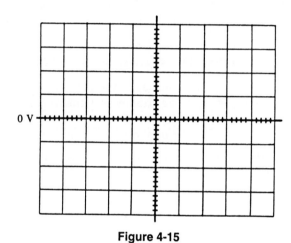

Figure 4-15

Vertical sensitivity = _____
Horizontal sensitivity = _____

d. Using the oscilloscope with the coupling switch in the DC position obtain the waveform for v_o and record on Fig. 4.16. Use the same sensitivities employed in Part 5(c) and be sure to preset the $v_o = 0$ V line using the GND position of the coupling switch. Label the maximum and

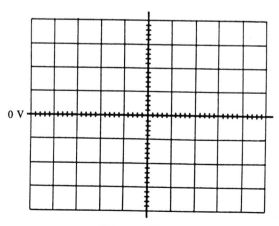

Figure 4-16

Procedure

minimum values of the waveform using the chosen vertical sensitivity.

How do the waveforms of Parts 5(**c**) and 5(**d**) compare?

e. Determine the DC level of the full-wave rectified waveform of Fig. 4.16.

(calculated) V_{DC} = _____

f. Measure the DC level of the output waveform using the DMM and calculate the percent difference between the measured and calculated values.

(measured) V_{DC} = _____

(% Difference) = _____

g. Replace diodes D_3 and D_4 by 2.2 kΩ resistors and forecast the appearance of the output voltage v_o including the effects of V_T for each diode. Sketch the waveform on Fig. 4.17 and label the magnitude of the maximum and minimum values. Record your choice of sensitivities below.

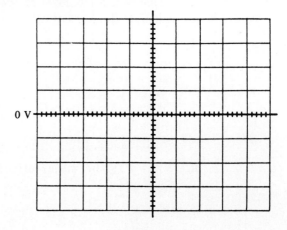

Figure 4-17

Vertical sensitivity = _____

Horizontal sensitivity = _____

h. Using the oscilloscope, obtain the waveform for v_o and reproduce on Fig. 4.18 indicating the maximum and minimum values. Use the same sensitivities as determined in Part 5(**g**).

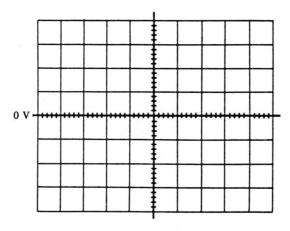

Figure 4-18

How do the waveforms of Fig. 4.17 and 4.18 compare?

i. Calculate the DC level of the waveform of Fig. 4.18.

(calculated) V_{DC} = _____

j. Measure the DC level of the output voltage using the DMM and calculate the percent difference.

(measured) V_{DC} = _____
(% Difference) = _____

k. What was the major effect of replacing the two diodes with resistors?

Procedure

Part 6. Full-Wave Center-Tapped Configuration

a. Construct the network of Fig. 4.19. Record the measured value of the resistor R.

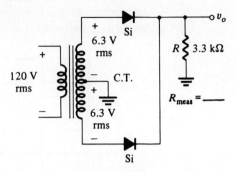

Figure 4-19

Measure each secondary voltage of the transformer with the DMM set on AC. Record below. Do they differ from the 6.3 V rating?

(measured) V_{rms} = _____
(measured) V_{rms} = _____

Using an average of the rms readings calculate the peak value of the secondary voltage.

(calculated) V_{peak} = _____

b. Using the V_T of Part 1 for each diode sketch the expected output waveform v_o on Fig. 4.20. Choose a vertical and horizontal sensitivity commensurate with the secondary voltage. Consult your oscilloscope to obtain a list of possibilities. Record your choice for each below.

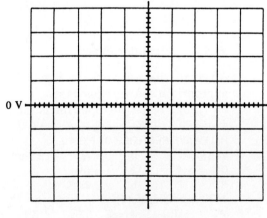

Figure 4-20

Vertical sensitivity = _____
Horizontal sensitivity = _____

c. Using the oscilloscope with the coupling switch in the DC position obtain the waveform for v_o and record on Fig. 4.21. Use the same sensitivities employed in Part 6(**b**) and be sure to preset the $v_o = 0$ V line using the GND position of the coupling switch. Label the maximum and minimum values of the waveform using the chosen vertical sensitivity.

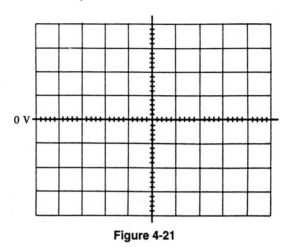

Figure 4-21

How do the waveforms of Figs. 4.20 and 4.21 compare?

d. Determine and compare the calculated and measured values of the DC level associated with v_o.

(calculated) = _____
(measured) = _____

Part 7. Computer Exercise

Analyze the network of Fig. 4.3 using PSpice, MicroCap II or other appropriate software package.
    For PSpice the input file is the following:

```
HALFWAVE RECTIFIER OF FIG 4.3
VS      1       0       AC      2.828V  0
D1      1       2       DI
R       2       0       2.2K
.MODEL DI D(IS=2E-15)
.AC     LIN     1       1KH     1KH
.OPTIONS NOPAGE
.PROBE
.END
```

Compare the results with those obtained in Part 2.

Name _____
Date _____
Instructor _____

EXPERIMENT 5

Clipping Circuits

OBJECTIVE

To become familiar with the function and operation of clippers.

EQUIPMENT REQUIRED

Instruments

Oscilloscope
DMM

Components

Resistors

(1) 2.2-kΩ

Diode

(1) Silicon
(1) Germanium

Supplies

(1) 1.5-V D cell and holder
Function generator

EQUIPMENT ISSUED

Item	Laboratory serial no.
Oscilloscope	
DMM	
Function generator	

RÉSUMÉ OF THEORY

The primary function of clippers is to "clip" away a portion of an applied alternating signal. The process is typically performed by a resistor-diode combination, but DC batteries are also incorporated to provide additional shifts or "cuts" of the applied voltage. The analysis of clippers with square-wave inputs is the easiest to perform since there are only two levels of input voltage to be concerned about. Each level can be treated as a DC input and the output voltage for the corresponding time time determined. For sinusoidal and triangular inputs, various instantaneous values can be treated as DC levels and the output level determined. Once a sufficient number of plot points for v_o have been determined, the output voltage can be sketched in total. Once the basic behavior of clippers is established, the effect of the placement of elements in various positions can be predicted and the analysis completed with less effort and less concern about accuracy.

PROCEDURE

Part 1. Threshold Voltage

Determine the threshold voltage for the silicon and germanium diodes using the diode-checking capability of the DMM or a curve tracer. Round off to hundredths place when recording in the designated space below. If the diode-checking capability or curve tracer is unavailable assume $V_T = 0.7$ V for the silicon diode and 0.3 V for the germanium diode.

$V_T(\text{Si}) =$ _____

$V_T(\text{Ge}) =$ _____

Part 2. Parallel Clippers

a. Construct the clipping network of Fig. 5.1. Record the measured resistance value and voltage of the D cell. Note that the input is an 8 V_{p-p} square wave at a frequency of 1000 Hz.

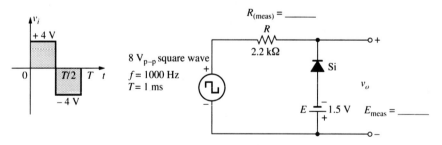

Figure 5-1

b. Using the measured values of R, E, and V_T calculate the voltage V_o when the applied square wave is +4 V. That is, for the interval when the input is +4 V what is the level of V_o? Show all the details of your calculations to determine V_o.

(calculated) $V_o =$ _____

Procedure

c. Repeat Part 2(**b**) when the applied square wave is –4 V.

(calculated) $V_o =$ _____

d. Using the results of Parts 2(**b**) and 2(**c**) sketch the expected waveform for v_o using the horizontal axis of Fig. 5.2 as the $V_o = 0$ V line. Use a vertical sensitivity of 1V/cm and a horizontal sensitivity of 0.2 ms/cm.

Calculated:

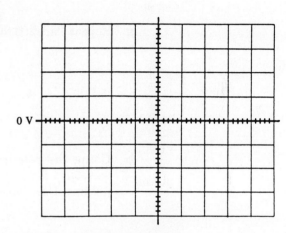

Figure 5-2

e. Using the sensitivities provided in Part 2(**d**) set the input square wave and record v_o on Figure 5.3 using the oscilloscope. Be sure to preset the $V_o = 0$ V line using the GND position of the coupling switch (and the DC position to view the waveform).

Measured:

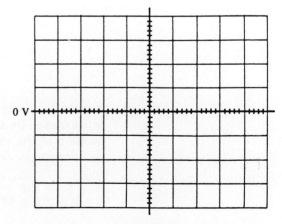

Figure 5-3

How does the waveform of Fig. 5.3 compare with the predicted result of Fig. 5.2?

f. Reverse the battery of Fig. 5.1 and using the measured values of R, E, and V_T calculate the level of V_o for the time interval when $V_i = +4$ V.

(calculated) $V_o = $ _____

g. Repeat Part 2(**f**) for the time interval when $V_i = -4$ V.

(calculated) $V_o = $ _____

h. Using the results of Parts 2(**f**) and 2(**g**) sketch the expected waveform for v_o using the horizontal axis of Fig. 5.4 as the $V_o = 0$ V line. Use the same sensitivities provided in part 2(**d**).

Calculated:

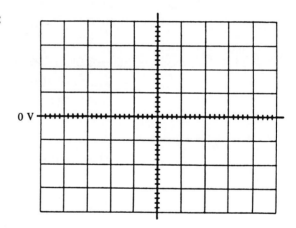

Figure 5-4

i. Set the input square wave and record v_o on Fig. 5.5 using the oscilloscope. Be sure to preset the $V_o = 0$ V line using the GND position of the coupling switch (and the DC position to view the waveform)

How does the waveform of Fig. 5.4 compare with the predicted result of Fig. 5.5?

Procedure

Measured:

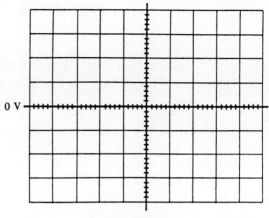

Figure 5-5

Part 3. Parallel Clippers (continued)

a. Construct the network of Fig. 5.6. Record the measured value of the resistance. Note that the input is now a 4 V_{p-p} square wave at $f = 1000$ Hz.

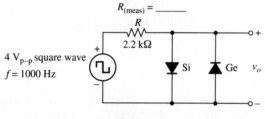

Figure 5-6

b. Using the levels of V_T determined in Part 1 calculate the level of V_o for the time interval when $V_i = +2$ V.

(calculated) V_o = _____

c. Repeat Part 3(b) for the time interval when $V_i = -2$ V.

(calculated) V_o = _____

d. Using the results of Parts 3(b) and 3(c) sketch the expected waveform for v_o using the horizontal axis of Fig. 5.7 as the $V_o = 0$ V line. Insert your chosen vertical and horizontal sensitivities below:

Vertical sensitivity = _____
Horizontal sensitivity = _____

Calculated:

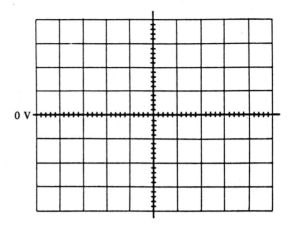

Figure 5-7

e. Using the sensitivities chosen in part 3(**d**) set the input square wave and record v_o on Fig. 5.8 using the oscilloscope. Be sure to preset the $V_o = 0$ V line using the GND position of the coupling switch (and the DC position to view the waveform).

Measured:

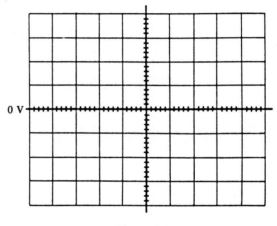

Figure 5-8

How does the waveform of Fig. 5.8 compare with predicted result of Fig. 5.7?

Part 4. Parallel Clippers (Sinusoidal Input)

a. Rebuild the circuit of Fig. 5.1 but change the input signal to an 8 $V_{p\text{-}p}$ sinusoidal signal with the same frequency (1000 Hz).

b. Using the results of Part 2 and any other analysis technique sketch the expected output waveform for v_o on Fig. 5.9. In particular find V_o when the applied signal is at its positive and negative peak and zero volts. Also list the chosen vertical and horizontal sensitivities below:

Calculated:

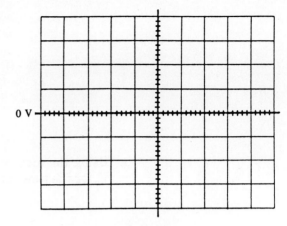

Figure 5-9

(calculated) V_o when $V_i = +4$ V is = _____
(calculated) V_o when $V_i = -4$ V is = _____
(calculated) V_o when $V_i = 0$ V is = _____
Vertical sensitivity = _____
Horizontal sensitivity = _____

c. Using the sensitivities chosen in Part 4(b) set the input sinusoidal waveform and record v_o on Fig. 5.10 using the oscilloscope. Be sure to preset the $V_o = 0$ V line using the GND position of the coupling switch.

Measured:

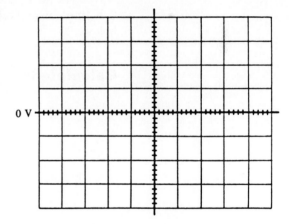

Figure 5-10

How does the waveform of Fig. 5.10 compare with the predicted result of Fig. 5.9?

Part 5. Series Clippers

a. Construct the circuit of Fig. 5.11. Record the measured resistance value and the DC level of the D cell. The applied signal is an 8 $V_{p\text{-}p}$ square wave at a frequency of 1000 Hz.

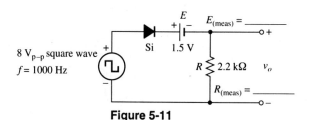

Figure 5-11

b. Using the measured values of R, E, and V_T calculate the voltage V_o for the time interval when $V_i = +4$ V.

(calculated) $V_o =$ _____

c. Repeat Part 5(**b**) for the time interval when $V_i = -4$ V.

(calculated) $V_o =$ _____

d. Using the results of Parts 5(**b**) and 5(**c**) sketch the expected waveform for v_o using the horizontal axis of Fig. 5.12 as the $V_o = 0$ V line. Insert your chosen vertical and horizontal sensitivities below:

Calculated:

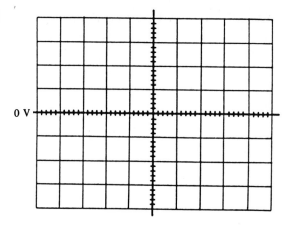

Figure 5-12

Vertical sensitivity = _____
Horizontal sensitivity = _____

e. Using the sensitivities chosen in Part 5(**d**) set the input square wave and record v_o on Fig. 5.13 using the oscilloscope. Be sure to preset the $V_o = 0$ V line using the GND position of the coupling switch (and the DC position to view the waveform).

Measured:

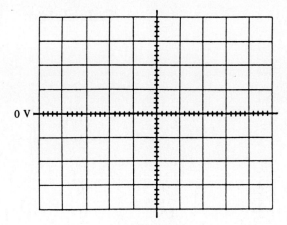

Figure 5-13

How does the waveform of Fig. 5.13 compare with the predicted result of Part 5(**d**)?

f. Reverse the battery of Fig. 5.11 and using the measured values of R, E, and V_T calculate the level of V_o for the time interval when $V_i = +4$ V.

(calculated) $V_o =$ _____

g. Repeat Part 5(**f**) for the time interval when $V_i = -4$ V.

(calculated) $V_o =$ _____

h. Using the results of Parts 5(**f**) and 5(**g**) sketch the expected waveform for v_o using the horizontal axis of Fig. 5.14 as the $V_o = 0$ V line. Use the following sensitivities:

Vertical: 2 V/cm

Horizontal: 0.2 ms/cm

Calculated:

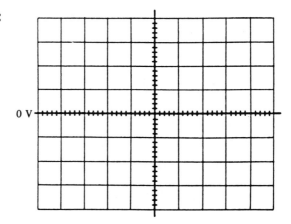

Figure 5-14

i. Using the sensitivities provided in Part 5(**h**) set the input square wave and record v_o on Fig. 5.15 using the oscilloscope. Be sure to preset the $V_o = 0$ V line using the GND position of the coupling switch (and the DC position to view the waveform).

Measured:

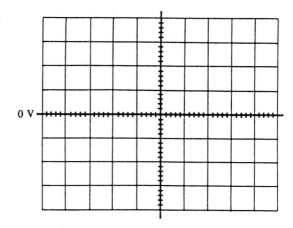

Figure 5-15

How does the waveform of Fig. 5.15 compare with the predicted pattern of Fig. 5.14?

Part 6. Series Clippers (Sinusoidal Input)

a. Rebuild the circuit of Fig. 5.11 but change the input signal to an 8 $V_{p\text{-}p}$ sinusoidal signal with the same frequency (1000 Hz).

b. Using the results of Part 5 and any other analysis technique sketch the expected output waveform for v_o on Fig. 5.16. In particular, find V_o when the applied signal is at its positive and negative peak and zero volts. Use a vertical sensitivity of 1 V/cm and a horizontal sensitivity of 0.2 ms/cm.

Calculated:

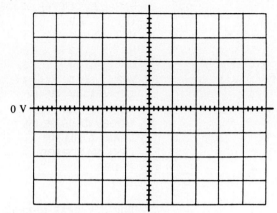

Figure 5-16

(calculated) V_o when $V_i = +4$ V is = _____
(calculated) V_o when $V_i = -4$ V is = _____
(calculated) V_o when $V_i = 0$ V is = _____

c. Using the sensitivities provided in Part 6(b) set the input sinusoidal waveform and record v_o on Fig. 5.17 using the oscilloscope. Be sure to preset the $V_o = 0$ V line using the GND position of the coupling switch.

Measured:

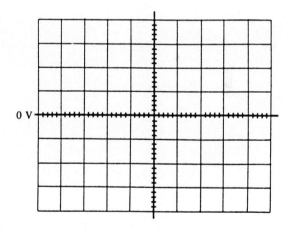

Figure 5-17

How does the waveform of Fig. 5.17 compare with the predicted result of Fig. 5.16?

Part 7. Computer Exercise

Analyze the network of Fig. 5.1 using PSpice, MicroCap II, or other appropriate software package.

For PSpice the input file is the following:

```
PARALLEL CLIPPER OF FIGURE 5.1
VE1       0      1       4V
VS        2      1       PULSE (0V 8V 0S 1NS 1NS 0.5MS 1MS)
R         2      3       2.2K
D1        4      3       DI
VE2       0      4       1.5V
.DC  VE1  4V  4V  1V
.DC  VE2  1.5V  1.5V  1V
.MODEL  DI (IS=2E-15)
.TRAN  0.05M  1M
.OPTIONS NOPAGE
.PROBE
.END
```

The 8 V peak-to-peak square wave is formed by the second and third lines of the input file. The first two entries within the parentheses of the PULSE statement specify that the pulse will rise from 0 V to 8 V. The next entry specifies zero seconds for the delay time while the next two time intervals specify the rise and fall times respectively. The 0.5 ms entry is the pulse width for the 8 V level and the 1 ms is the period of the pulse. Recall that the period of a 1 kHz signal is 1 ms. The source VE1 has a polarity opposite to that of VS and will establish the square wave pulse that extends from –4 V to +4 V. The .TRAN statement will generate a data point for the PROBE run to follow every 0.5 ms or every 50 μs up to the period of 1 ms.

Compare the computer generated result with that of Part 2.

Name	
Date	
Instructor	

EXPERIMENT 6

Clamping Circuits

OBJECTIVE

To become familiar with the function and operation of clampers.

EQUIPMENT REQUIRED

Instruments

Oscilloscope
DMM

Components

Resistors

(1) 100-Ω
(1) 1-kΩ
(1) 100-kΩ

Diode

(1) Silicon

Capacitor

(1) 1 μF

Supplies

(1) 1.5-V D cell and holder
Function generator

EQUIPMENT ISSUED

Item	Laboratory serial no.
Oscilloscope	
DMM	
Function generator	

RÉSUMÉ OF THEORY

Clampers are designed to "clamp" an alternating input signal to a specific level without altering the peak-to-peak characteristics of the waveform. Clampers are easily distinguished from clippers in that they include a capacitive element. A typical clamper will include a capacitor, diode, and resistor with some also having a DC battery. The best approach to the analysis of clampers is to use a step-by-step approach. The first step should be an examination of the network for that part of the input signal that forward biases the diode. Choosing this part of the input signal will save time and probably avoid some unnecessary confusion. With the diode forward biased the voltage across the capacitor and across the output terminals can be determined. For the rest of the analysis it is then assumed that the capacitor will hold on to the charge and voltage level established during this interval of the input signal. The next part of the input signal can then be analyzed to determine the effect of the stored voltage across the capacitor and the open-circuit state of the diode.

The analysis of a clamper can be quickly checked by simply noting whether the peak-to-peak voltage of the output signal is the same as the peak-to-peak voltage of the applied signal. This check is not sufficient to be sure the entire analysis was correct but it is a characteristic of clampers that must be satisfied.

PROCEDURE

Part 1. Threshold voltage

Determine the threshold voltage for the silicon diode using the diode-checking capability of the DMM or a curve tracer. If either approach is unavailable assume $V_T = 0.7$ V.

$V_T =$ _____

Part 2. Clampers (R, C, diode combination)

a. Construct the network of Fig. 6.1 and record the measured value of R.

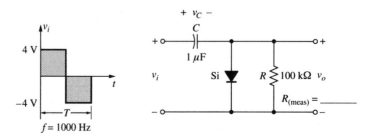

$R_{(meas)} =$ _____

Figure 6-1

Procedure

b. Using the value of V_T from Part 1 calculate V_C and V_o for the interval of v_i that causes the diode to be in the "on" state.

(calculated) $V_C =$ _____

(calculated) $V_o =$ _____

c. Using the results of Part 2(**b**) calculate the level of V_o after v_i switches to the other level and turns the diode "off".

(calculated) $V_o =$ _____

d. Using the results of Parts 2(**b**) and 2(**c**) sketch the expected waveform for V_o in Fig. 6.2 for one full cycle of V_i. Use the horizontal center axis as the $V_o = 0$ V line. Record the chosen vertical and horizontal sensitivities below:

Calculated:

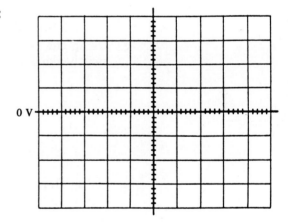

Figure 6-2

Vertical sensitivity = _____

Horizontal sensitivity = _____

e. Using the sensitivities of Part 2(**b**) use the oscilloscope to view the output waveform v_o. Be sure to preset the $V_o = 0$ V line on the screen using the GND position of the coupling switch (and the DC position to view the waveform). Record the resulting waveform on Fig. 6.3.

How does the waveform of Fig. 6.3 compare with the expected waveform of Fig. 6.2?

Measured:

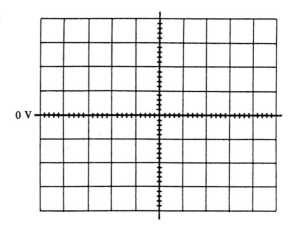

Figure 6-3

f. Reverse the diode of Fig. 6.1 and using the value of V_T from Part 1 determine the levels of V_C and V_o for the interval of v_i that causes the diode to be in the "on" state.

(calculated) $V_C =$ _____

(calculated) $V_o =$ _____

g. Using the results of Part 2(f) calculate the level of V_o after v_i switches to the other level and turns the diode "off."

(calculated) $V_o =$ _____

h. Using the results of Parts 2(f) and 2(g) sketch the expected waveform for v_o on Fig. 6.4. Use the horizontal axis as the $v_o = 0$ V line. Record the chosen vertical and horizontal sensitivities below:

Calculated:

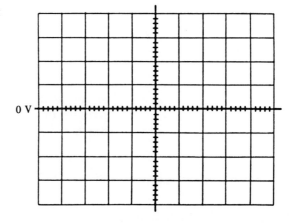

Figure 6-4

Procedure

Vertical sensitivity = _____
Horizontal sensitivity = _____

i. Using the sensitivities of Part 2(**h**) use the oscilloscope to view the output waveform v_o. Be sure to preset the $V_o = 0$ V line on the screen using the GND position of the coupling switch (and the DC position to view the waveform). Record the resulting waveform on Fig. 6.5.

Measured:

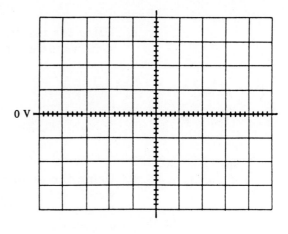

Figure 6-5

How does the waveform of Fig. 6.5 compare with the expected waveform of Fig. 6.4?

Part 3. Clampers with a DC battery

a. Construct the network of Fig. 6.6 and record the measured values of R and E.

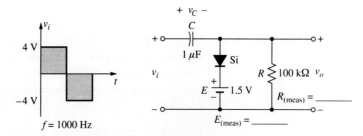

Figure 6-6

b. Using the value of V_T from Part 1 calculate V_C and v_o for that interval of v_i that causes the diode to be in the "on" state.

(calculated) V_C = _____

(calculated) V_o = _____

c. Using the results of Part 3(b) calculate the level of v_o after v_i switches to the other level and turns the diode "off."

(calculated) V_o = _____

d. Using the results of Parts 3(b) and 3(c) sketch the expected waveform for v_o on Fig. 6.7. Use the horizontal center axis as the $V_o = 0$ V line. Record the chosen vertical and horizontal sensitivities below:

Calculated:

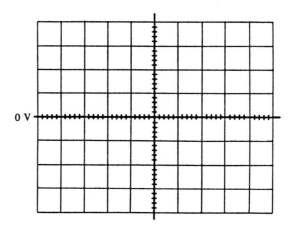

Figure 6-7

Vertical sensitivity = _____

Horizontal sensitivity = _____

e. Using the sensitivities of Part 3(d) use the oscilloscope to view the output waveform v_o. Be sure to preset the $V_o = 0$ V line on the screen using the GND position of the coupling switch (and the DC position to view the waveform). Record the resulting waveform on Fig. 6.8.

How does the waveform of Fig. 6.8 compare with the expected waveform of Fig. 6.7?

Procedure

Measured:

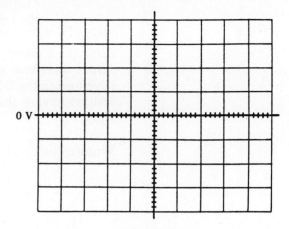

Figure 6-8

f. Reverse the diode of Fig. 6.6 and using the value of V_T from Part 1 calculate the levels of V_C and V_o for that interval of the input voltage v_i that causes the diode to be in the "on" state.

(calculated) $V_C =$ _____
(calculated) $V_o =$ _____

g. Using the results of Part 3(f) calculate the level of V_o after v_i switches to the other level and turns the diode "off."

(calculated) $V_o =$ _____

h. Using the results of Parts 3(f) and 3(g) sketch the expected waveform for v_o on Fig. 6.9. Use the horizontal center axis as the $V_o = 0$ V line. Record the chosen vertical and horizontal sensitivities below:

Calculated:

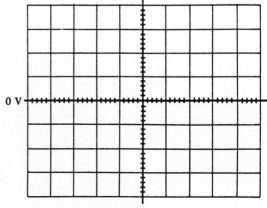

Figure 6-9

Vertical sensitivity = _____
Horizontal sensitivity = _____

i. Using the sensitivities of Part 3(**h**) use the oscilloscope to view the output waveform v_o. Be sure to preset the $V_o = 0$ V line on the screen using the GND position of the coupling switch (and the DC position to view the waveform). Record the resulting waveform on Fig. 6.10.

Measured:

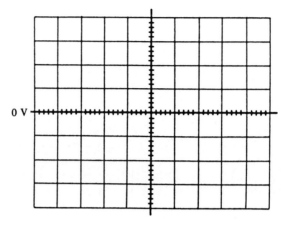

Figure 6-10

How does the waveform of Fig. 6.10 compare with the expected waveform of Fig. 6.9?

Part 4. Clampers (sinusoidal input)

a. Reconstruct the network of Fig. 6.1 but change the input signal to an 8 V_{p-p} sinusoidal signal with the same frequency (1000 Hz).

b. Using the results of Parts 1 and 2 and any other analysis technique at your disposal, sketch the expected output waveform for v_o on Fig. 6.11. In particular find v_o when v_i is its positive and negative peak value and when $v_i = 0$ V. Record the chosen vertical and horizontal sensitivities below:

Calculated:

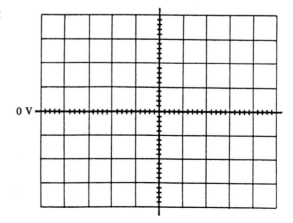

Figure 6-11

Procedure

(calculated) V_o when V_i = +4 V is _____
(calculated) V_o when V_i = −4 V is _____
(calculated) V_o when V_i = 0 V is _____
Vertical sensitivity = _____
Horizontal sensitivity = _____

c. Using the sensitivities of Part 4(**b**) use the oscilloscope to view the output waveform v_o. Be sure to preset the V_o = 0 V line on the screen using the GND position of the coupling switch (and the DC position to view the waveform). Record the resulting waveform on Fig. 6.12.

Measured:

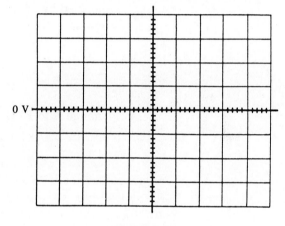

Figure 6-12

How does the waveform of Fig. 6.12 compare with the expected waveform of Fig. 6.11?

Part 5. Clampers (Effect of *R*)

a. Determine the time constant ($\tau = RC$) for the network of Fig. 6.1 for that interval of the input signal that causes the diode to assume the "off" state and be approximated by an open-circuit.

(calculated) τ = _____

b. Calculate the period of the applied signal and then determine half the period to correspond with the time interval that the diode is in the "off" state during the first cycle of the applied signal.

(calculated) T = _____
(calculated) $T/2$ = _____

c. The discharge period of an RC network is about 5τ. Calculate the time interval established by 5τ using the result of Part 5(a) and compare to $T/2$ calculated in Part 5(b).

(calculated) 5τ = _____

d. For good clamping action, why is it important for the time interval specified by 5τ to be much larger than $T/2$ of the applied signal?

e. Change R to 1 kΩ and calculate the new value of 5τ.

(calculated) 5τ = _____

f. How does the 5τ calculated in Part 5(e) compare to $T/2$ of the applied signal? How would you expect the new value of R to affect the output waveform v_o?

g. Set the input of Fig. 6.1 with R = 1 kΩ and record the resulting waveform on Fig. 6.13. Be sure to preset the V_o = 0 V line in the center of the screen using the GND position of the coupling switch and be sure to use the DC position to view the waveform. Insert the chosen vertical and horizontal sensitivities below:

Procedure

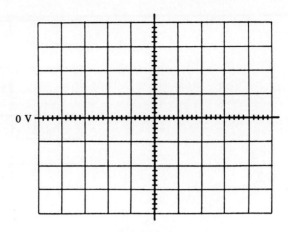

Figure 6-13

Vertical sensitivity = _____
Horizontal sensitivity = _____

h. Comment on the resulting waveform of Fig. 6.13. Is the distortion as you expected it to appear? Are you surprised by the positive and negative peaks? Why?

i. Change R to 100 Ω and calculate the new value of 5τ.

(calculated) 5τ = _____

j. How does the 5τ calculated in Part 5(i) compare to $T/2$ for the applied signal? What effect will the lower value of R have on the waveform of Fig. 6.13?

k. Set the input of Fig. 6.1 with R = 100 Ω and record the resulting waveform on Fig. 6.14. Be sure to preset the V_o = 0 V line using the coupling switch and use the DC position to view the waveform v_o. Insert the chosen vertical and horizontal sensitivities below:

Vertical sensitivity = _____
Horizontal sensitivity = _____

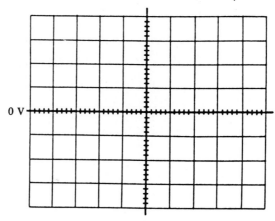

Figure 6-14

l. Comment on the resulting waveform of Fig. 6.14. Compare it to the waveform of Fig. 6.13 and the properly clamped waveform of Fig. 6.3.

m. Based on the results of Parts 5(a) through 5(l) establish a relationship between 5τ and the period of the waveform (T) that will insure that the output waveform has the same characteristics as the input. Note that the requested relationship is between 5τ and T and not $T/2$.

Part 6. Computer Exercise

a. Analyze the network of Fig. 6.1 using PSpice, MicroCap II, or other appropriate software package.

For PSpice the input file is the following:

```
CLAMPER OF FIG. 6.1
VE1   0    1    2V
VS    2    1    PULSE (0V 4V 0S 1NS 1NS 0.5MS 1MS)
C1    2    3    1UF
D1    3    0    DI
R1    3    0    100K
.DC   VE1  2V   2V   1V
.MODEL  DI (IS=2E-15)
.TRAN  0.05M  1M
.OPTIONS  NOPAGE
.PROBE
.END
```

Procedure

The square wave input is established in the same manner as described for Part 6 of Experiment 5.
Compare to the results of Part 1

    b. Repeat the analysis of Part 5(a) but with $R = 1$ kΩ. Comment on the resulting shape of the v_o curve using PROBE.

    c. Repeat the analysis of Part 5(a) but with $R = 100$ Ω. Comment on the resulting shape of the v_o curve using PROBE.

Name _____
Date _____
Instructor _____

EXPERIMENT 7

Light-Emitting and Zener Diodes

OBJECTIVE

To become familiar with the characteristics and use of a light-emitting diode (LED) and Zener diode.

EQUIPMENT REQUIRED

Instruments

DMM

Components

Resistors

(1) 100-Ω
(1) 220-Ω
(1) 330-Ω
(1) 2.2-kΩ
(1) 3.3-kΩ
(2) 1-kΩ

Diode

(1) Silicon
(1) LED
(1) Zener (10-V)

Supplies

DC power supply

EQUIPMENT ISSUED

Item	Laboratory serial no.
DMM	
DC power supply	

RÉSUMÉ OF THEORY

The light-emitting diode (LED) is, as the name implies, a diode that will give off visible light when sufficiently energized. In any forward-biased *p-n* junction there is, close to the junction, a recombination of holes and electrons. This recombination requires that the energy possessed by unbound free electrons be transferred to another state. In LED materials, such as gallium arsenide phosphide (GaAsP) or gallium phosphide (GaP), photons of light energy are emitted in sufficient numbers to create a visible light source—a process referred to as *electroluminescence*. For every LED there is a distinct forward voltage and current that will result in a bright, clear light, whether it be red, yellow, or green. The diode may, therefore, be forward biased, but until the distinct level of voltage and current are reached, the light may not be visible. In this experiment the characteristics of an LED will be plotted and the "firing" levels of voltage and current determined.

The Zener diode is a *p-n* junction device designed to take full advantage of the Zener breakdown region. Once the reverse-bias potential reaches the Zener region, the ideal Zener diode is assumed to have a fixed terminal voltage and zero internal resistance. All practical diodes have some internal resistance even though, typically, it is limited to 5 to 20 Ω. The internal resistance is the source of the variation in Zener voltage with current level. The experimental procedure will demonstrate the variation in terminal voltage for different loads and resulting current levels.

For most configurations, the state of the Zener diode can usually be determined simply by replacing the Zener diode with an open circuit and calculating the voltage across the resulting open circuit. If the open-circuit voltage equals or exceeds the Zener potential, the Zener diode is "on" and the Zener diode can be replaced by a DC supply equal to the Zener potential. Even though the open-circuit voltage may be greater than the Zener potential, the diode is still replaced by a supply equal to the Zener potential. The foregoing procedure is used to determine the state of the Zener diode. Once the Zener voltage is substituted, the remaining voltages and currents of the network can be determined.

PROCEDURE

Part 1. LED Characteristics

a. Construct the circuit of Fig. 7.1. Initially, set the supply to 0 V and record the measured value of the resistor R.

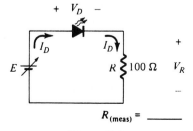

$R_{(meas)} = $ _____

Figure 7-1

b. Increase the supply voltage E until "first light" is noticed. Record the value of V_D and V_R using the DMM. Calculate the corresponding level of I_D using $I_D = V_R/R$ and the measured resistance value.

(measured) V_D = _____

(measured) V_R = _____

(calculated) I_D = _____

c. Continue to increase the supply voltage E until "good brightness" is first established. Don't overload (too much current) the circuit and possibly damage the LED by continuing to raise the voltage beyond this level. Record the values of V_D and V_R and calculate the corresponding level of I_D using $I_D = V_R/R$ and the measured resistance value.

(measured) V_D = _____

(measured) V_R = _____

(calculated) I_D = _____

d. Set the DC supply to the levels appearing in Table 7.1 and measure both V_D and V_R. Record the values of V_D and V_R in Table 7.1 and calculate the corresponding level of I_D using $I_D = V_R/R$ and the measured resistance value.

TABLE 7.1

E(V)	0	1	2	3	4	5	6
V_D(V)							
V_R(V)							
$I_D = V_R/R$ (mA)							

e. Using the data of Table 7.1 sketch the curve of I_D vs. V_D on the graph of Fig. 7.2. Choose an appropriate scale for both I_D and V_D.

f. Draw a light dashed horizontal line across the graph of Fig. 7.2 at the current I_D required for "good brightness." In addition, draw a light dashed vertical line the full height of Fig. 7.2 at the point of intersection between the curve and the light dashed horizontal line. The intersection of the vertical line with the horizontal axis should result in a level of V_D close to that measured in Part 1(**c**)

Shade in the region below the I_D line and to the left of the V_D line and label the region as the region to be avoided if "good brightness" is to be obtained. Label the remaining unshaded region of Fig. 7.2 as the region for "good brightness."

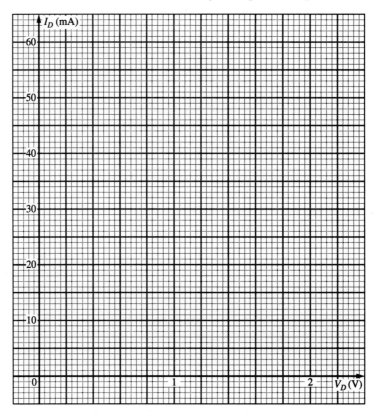

Figure 7-2

g. Construct the circuit of Fig. 7.3. Be sure that both diodes are connected properly and record the measured resistance value.

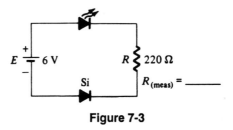

Figure 7-3

h. Do you expect the LED to burn brightly? Why?

i. Energize the network of Fig. 7.3 and verify your conclusion in step **h**.

j. Reverse the silicon diode of Fig. 7.3 and repeat step **h**.

Procedure

k. Repeat step **i**. If the LED is "on" with "good brightness" measure V_D and V_R and calculate the level of I_D. Find the intersection of I_D and V_D on the graph of Fig. 7.2. Is the intersection on the curve part of the "good brightness" region?

Part 2. Zener Diode Characteristics

a. Construct the circuit of Fig. 7.4. Initially, set the DC supply to 0 V and record the measured value of R.

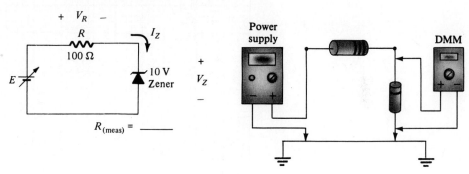

Figure 7-4

b. Set the DC supply (E) to the values appearing in Table 7.2 and measure both V_Z and V_R. You may have to use the millivolt range of your DMM for low values of V_Z and V_R.

TABLE 7.2

E(V)	0	1	2	3	4	5	6	7	8	9	10	11	12	13	14	15
V_Z(V)																
V_R(V)																
$I_Z = V_R/R_{meas}$ (mA)																

c. Calculate the Zener current I_Z in mA at each level of E using Ohm's law as indicated in the last row of Table 7.2 and complete the table.

d. This step will develop the characteristic curve for the Zener diode. Since the Zener region is in the third quadrant of a complete diode characteristic curve place a minus sign in front of each level of I_Z and V_Z for each data point. With this convention in mind plot the data of Table 7.2 on the graph of Fig. 7.5. Choose an appropriate scale for I_Z and V_Z as determined by the range of values for each parameter.

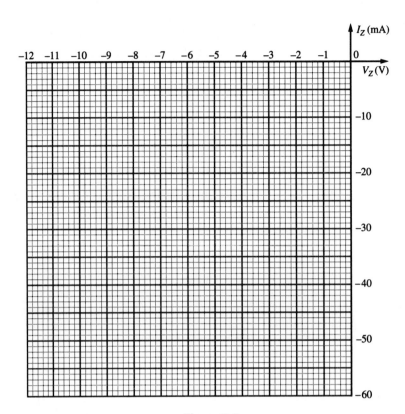

Figure 7-5

e. For the range of measurable current I_Z in the linear (straight line) region that drops from the V_Z axis, what is the average value of V_Z?

  In other words, for all practical purposes, what is V_Z for this Zener diode?

(approximated) V_Z = _____

f. For the range of measurable current I_Z in the linear region that drops from the V_Z axis, estimate the average resistance of the Zener diode using $r_{av} = \Delta V_Z/\Delta I_Z$, where ΔV_Z is the change in Zener voltage for the corresponding change in Zener current. Choose an interval of at least 20 V on the linear region of the curve. If necessary, use the data of Table 7.2. Show all work.

(calculated) R_Z = _____

g. Using the results of steps **e** and **f**, establish the Zener diode equivalent circuit of Fig. 7.6 for the "on" linear region. That is, insert the values of R_Z and V_Z.

R_Z = _____ V_Z = _____

Figure 7-6

Procedure

h. For the region from V_Z and $I_Z = 0$ to the point where the characteristic curve drops sharply from the V_Z axis calculate the resistance of the Zener diode using the equation $r = \Delta V_Z / \Delta I_Z$. Choose $\Delta V_Z = V_Z - 0\text{ V} = V_Z$ and substitute the resulting change in current (ΔI_Z) for this interval.

(calculated) $R_Z =$ _____

Is the calculated level the level you expected for the region in which the Zener diode is "off"? What would be an appropriate approximation for the Zener diode in this region?

Part 3. Zener Diode Regulation

a. Construct the network of Fig. 7.7. Record the measured value of each resistor.

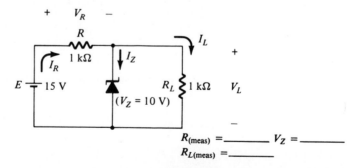

$R_{(meas)} =$ _____ $V_Z =$ _____
$R_{L(meas)} =$ _____

Figure 7-7

b. Determine whether the Zener diode of Fig. 7.7 is in the "on" state, that is, operating in the Zener breakdown region. Use the measured resistor values and the V_Z determined in Part 2(e).

Ignore the effects of R_Z in your calculations. For the diode in the "on" state calculate the expected values of V_L, V_R, I_R, I_L and I_Z. Show all calculations.

(calculated) V_L = _____
(calculated) V_R = _____
(calculated) I_R = _____
(calculated) I_L = _____
(calculated) I_Z = _____

c. Energize the network of Fig. 7.7 and measure V_L and V_R. Using these values calculate the levels of I_R, I_L and I_Z.

(measured) V_L = _____
(measured) V_R = _____
(calculated) I_R = _____
(calculated) I_L = _____
(calculated) I_Z = _____

How do the results of steps 3(b) and 3(c) compare?

d. Change R_L to 3.3 kΩ and repeat step b. That is, calculate the expected levels of V_L, V_R, I_R, I_L, and I_Z using measured resistor values and the V_Z determined in step 2(e).

Procedure

(calculated) V_L = _____
(calculated) V_R = _____
(calculated) I_R = _____
(calculated) I_L = _____
(calculated) I_Z = _____

e. Energize the network of Fig. 7.7 with R_L = 3.3 kΩ and measure V_L and V_R. Using these values calculate the levels of I_R, I_L and I_Z.

(measured) V_L = _____
(measured) V_R = _____
(calculated) I_R = _____
(calculated) I_L = _____
(calculated) I_Z = _____

How do the results of steps 3(**d**) and 3(**e**) compare?

f. Using the measured resistor values and V_Z determined from step 2(**e**), determine the minimum value of R_L required to insure that the Zener diode is in the "on" state.

(calculated) $R_{L_{min}}$ = _____

g. Based on the results of step 3(**f**), will a load resistor of 2.2 kΩ place the Zener diode of Fig. 7.7 in the "on" state?

Insert R_L = 2.2 kΩ into Fig. 7.7 and measure V_L.
(measured) V_L = _____

Are the conclusions of steps 3(**f**) and 3(**g**) verified?

Part 4. LED-Zener Diode Combination

a. In this part of the experiment we will determine the minimum supply voltage necessary to turn on (good brightness) the LED and the Zener diode of Fig. 7.8. The LED will reveal when the Zener diode is "on" and the required supply voltage will be the minimum value that can be applied if the Zener diode is to be used to regulate the voltage V_L.

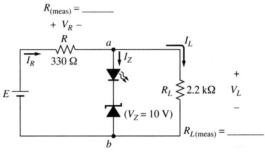

Figure 7-8

b. Refer to step 1(**c**) and record the level of V_D and I_D that resulted in a "good brightness" level for the LED.

$V_D =$ _____

$I_D =$ _____

Refer to step 2(**e**) and record the level of V_Z for your Zener diode.

$V_Z =$ _____

Using the above data determine the total voltage necessary to turn both the LED diode "on" in Fig. 7.8. That is, determine the required voltage from point a to b.

(calculated) $V_{ab} =$ _____

c. Using the result of step 4(**b**) calculate the voltage V_L and resulting current I_L. Use measured resistor values.

(calculated) $V_L =$ _____

(calculated) $I_L =$ _____

Procedure

d. Calculate I_R from $I_R = I_L + I_Z = I_L + I_D$ using the level of I_D from part 4(**b**). Then calculate the voltage V_R using Ohm's law.

(calculated) $I_R =$ _____
(calculated) $V_R =$ _____

e. Using Kirchhoff's voltage law calculate the required supply voltage E to turn on the Zener diode and establish "good brightness" by the LED. Use measured resistor values.

(calculated) $E =$ _____

f. Turn on the supply of Fig. 7.8 and increase the voltage E until the LED has "good brightness." Record the required level of E below:

(measured) $E =$ _____

How does the level calculated in step (**e**) compare with the measured value?

g. Measure the voltage V_D and compare with the level listed in step (**b**).

(measured) $V_D =$ _____

Measure the voltage V_Z and compare with the level listed in step (**b**).

(measured) $V_Z =$ _____

Name _____
Date _____
Instructor _____

EXPERIMENT 8
Bipolar Junction Transistor (BJT) Characteristics

OBJECTIVE

1. To determine transistor type (npn, pnp), terminals, and material using a digital multimeter (DMM).
2. To graph the collector characteristics of a transistor using experimental methods and a curve tracer.
3. To determine the value of the alpha and beta ratios of a transistor.

EQUIPMENT REQUIRED

Instruments

DMM
Curve Tracer (if available)

Components

Resistors

(1) 1 kΩ
(1) 330-kΩ
(1) 5-kΩ potentiometer
(1) 1-MΩ potentiometer

Transistors

(1) 2N3904 (or equivalent)
(1) Transistor without terminal identification

Supplies

DC power supply

EQUIPMENT ISSUED

Item	Laboratory serial no.
DMM	
Curve tracer	
DC power supply	

RÉSUMÉ OF THEORY

Bipolar transistors are made of either silicon (Si) or germanium (Ge). Their structure consists of two layers of *n*-type material separated by a layer of *p*-type material (*npn*), or of two layers of *p*-material separated by a layer of *n*-material (*pnp*). In either case, the center layer forms the base of the transistor, while the external layers form the collector and the emitter of the transistor. It is this structure that determines the polarities of any voltages applied and the direction of the electron or conventional current flow. With regard to the latter, the arrow at the emitter terminal of the transistor symbol for either type of transistor points in the direction of conventional current flow and thus provides a useful reference (Fig. 8.2). One part of this experiment will demonstrate how you can determine the type of transistor, its material, and identify its three terminals.

The relationships between the voltages and the currents associated with a bipolar junction transistor under various operating conditions determine its performance. These relationships are collectively known as the characteristics of the transistor. As such, they are published by the manufacturer of a given transistor in a specification sheet. It is one of the objectives of this laboratory experiment to experimentally measure these characteristics and to compare them to their published values.

PROCEDURE

Part 1. Determination of the Transistor's Type, Terminals, and Material

The following procedure will determine the type of a transistor, the terminals of a transistor, and the material from which it is made. The procedure will utilize the diode testing scale found on many modern multimeters. If no such scale is available, the resistance scales of the meter may be used.

    a. Label the transistor terminals of Fig. 8.1 as 1, 2, and 3. Use the transistor without terminal identification for this part of the experiment.

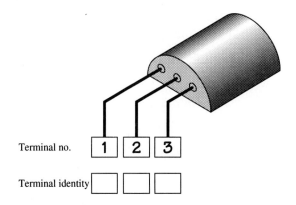

Figure 8-1 Determination of the identities of BJT leads.

Procedure

b. Set the selector switch of the multimeter to the diode scale (or to the 2 kΩ range if the diode scale is unavailable).

c. Connect the positive lead of the meter to terminal 1 and the negative lead to terminal 2; record your reading in Table 8.1.

TABLE 8.1

Step	Meterleads connected to BJT		Diode check reading
	Positive	Negative	(or highest resistance range)
c	1	2	
d	2	1	
e	1	3	
f	3	1	
g	2	3	
h	3	2	

d. Reverse the leads and record your reading.

e. Connect the positive lead to terminal 1 and the negative lead to terminal 3; record you reading.

f. Reverse the leads and record your reading.

g. Connect the positive lead to terminal 2 and the negative lead to terminal 3; record your reading.

h. Reverse the leads and record your reading.

i. The meter readings between two of the terminals will read high (O.L. or higher resistance) regardless of the polarity of the meter leads connected. Neither of these two terminals will be the base. Based on the above, record the number of the base terminal in Table 8.2.

TABLE 8.2

Part 1 (i):	Base terminal	
Part 1 (j):	Transistor type	
Part 1 (k):	Collector terminal	
Part 1 (k):	Emitter terminal	
Part 1 (l):	Transistor material	

j. Connect the negative lead to the base terminal and the positive lead to either of the other terminals. If the meter reading is low (approximately 0.7 V for Si and 0.3 V for Ge or lower resistance), the transistor type is *pnp;* go to step **k**(1). If the reading is high, the transistor type is *npn*; go to step **k**(2).

k. (1) For *pnp* type, connect the negative lead to the base terminal and the positive lead alternately to either of the other two terminals. The lower of the two readings obtained indicates that the base and collector are connected; thus the other terminal is the emitter. Record the terminals in Table 8.2.

(2) For *npn* type, connect the positive lead to the base terminal and the negative lead alternately to either of the other two terminals. The lower of the two readings obtained indicates that the base and collector are connected; thus the other terminal is the emitter. Record the terminals in Table 8.2.

l. If the readings in either (1) or (2) of Part 1(k) were approximately 700 mV, the transistor material is silicon. If the readings were approximately 300 mV, the material is germanium. If the meter does not have a diode testing scale, the material cannot be determined directly. Record the type of material in Table 8.2.

Part 2. The Collector Characteristics

a. Construct the network of Fig. 8.2.

b. Set the voltage V_{R_B} to 3.3 V by varying the 1 MΩ potentiometer. This adjustment will set $I_B = V_{R_B}/R_B$ to 10 μA as indicated in Table 8.3.

c. Then set V_{CE} to 2 V by varying the 5 kΩ potentiometer as required by the first line of Table 8.3.

d. Record the voltages V_{R_C} and V_{BE} in Table 8.3.

e. Vary the 5 kΩ potentiometer to increase V_{CE} from 2 V to the values appearing in Table 8.3. Note that I_B is maintained at 10 μA for the range of V_{CE} levels.

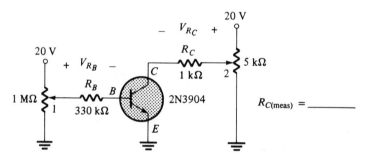

Figure 8-2 Circuit to determine the characteristics of a BJT.

f. For each value of V_{CE} measure and record V_{R_C} and V_{BE}. Use the mV scale for V_{BE}.

g. Repeat steps (b) through (f) for all values of V_{R_B} indicated in Table 8.3. Each value of V_{R_B} will establish a different level of I_B for the sequence of V_{CE} values.

h. After all data have been obtained, compute the values of I_C from $I_C = V_{RC}/R_C$ and I_E from $I_E = I_C + I_B$. Use the measured resistor value for R_C.

i. Using the data of Table 8.3, plot the collector characteristics of the transistor on the graph of Fig. 8.3. That is, plot I_C versus V_{CE} for the various values of I_B. Choose an appropriate scale for I_C and label each I_B curve.

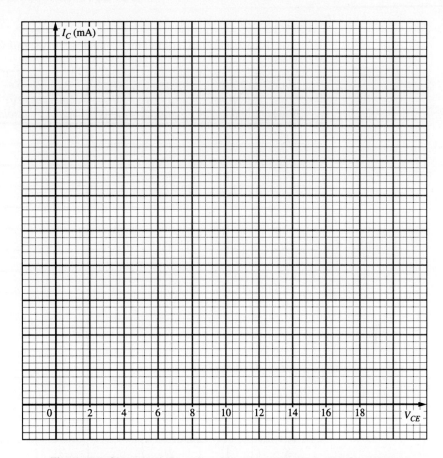

Figure 8.3 Characteristic curves from the experimental data of Part 2.

Part 3. Variation of α and β

a. For each line of Table 8.3 calculate the corresponding levels of α and β using $\alpha = I_C/I_E$ and $\beta = I_C/I_B$ and complete the Table.

b. Is there a significant variation in α and β from one region of the characteristics to another?

TABLE 8.3
Data for Construction of Transistor Collector Curve and Calculations of Transistor Parameters

V_{RB} (V) (meas)	I_B (μA) (calc)	V_{CE} (V) (meas)	V_{RC} (V) (meas)	I_C (mA) (calc)	V_{BE} (V) (meas)	I_E (mA) (calc)	α (calc)	β (calc)
↑	↑	2						
│	│	4						
│	│	6						
3.3	10	8						
│	│	10						
│	│	12						
│	│	14						
↓	↓	16						
↑	↑	2						
│	│	4						
│	│	6						
6.6	20	8						
│	│	10						
│	│	12						
↓	↓	14						
↑	↑	2						
│	│	4						
9.9	30	6						
│	│	8						
↓	↓	10						
↑	↑	2						
13.2	40	4						
│	│	6						
↓	↓	8						
↑	↑	2						
16.5	50	4						
↓	↓	6						

In which region are the largest values of β found? Specify using the relative levels of V_{CE} and I_C.

In which region are the smallest values of β found? Specify using the relative levels of V_{CE} and I_C.

Procedure

c. Find the largest and smallest levels of β and mark their locations on the plot of Fig. 8.3 using the notation $β_{max}$ and $β_{min}$.

d. In general, did β increase or decrease with increase in I_C?

e. In general, did β increase or decrease with increase in V_{CE}? Was the effect of V_{CE} on β greater or less than the effect of I_C?

Part 4. Determination of the Characteristics of a Transistor

Using a Comercial Curve Tracer

a. If available, use a curve tracer to obtain a set of collector characteristics for the 2N3904 transistor. Use the 10 μA step function for I_B and choose a scale for V_{CE} and I_C that matches the scales appearing in the plot of Fig. 8.3.

b. Reproduce the characteristics obtained on the graph of Fig. 8.4. Be sure to label each I_B curve and include the scale for each axis.

c. Compare the characteristics to those obtained in Part 2. Be specific in describing the differences between the two sets of characteristics.

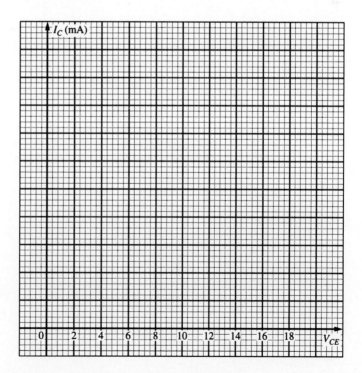

Figure 8-4 Characteristic curves obtained from commercial curve tracer.

Exercises

1. Find the average value of β using the data of Table 8.3. That is, find the sum of the β values and divide by the number of values.

(calculated) $\beta_{(av)}$ = _____

Where on the characteristics did the average value of β typically occur?

Is it reasonable to use this value of β for the transistor for most applications?

2. Determine the average value of V_{BE} using the data of Table 8.3. As in Exercise 1 find the sum of the V_{BE} values and divide by the number of values.

(calculated) $V_{BE\ (av)}$ = _____

Is it reasonable to use the 0.7 V level in the analysis of BJT transistor networks where the actual value is unknown?

Exercises

3. Careful inspection of the collector curves obtained by experimental measurements and by the curve tracer reveal that the slopes of constant base current are increasing positively (steeper) for higher base currents and higher levels of collector current. What is the effect of the increasing slope of the constant base current lines on the beta of the transistor?

Does the data of Table 8.3 substantiate the above conclusion?

If all the lines of constant base current were horizontal, what would be the effect on the beta ratio determined at any point on a particular base current curve?

If all the lines of constant base current were horizontal and equally spaced what would be the effect on the beta ratio determined anywhere on the characteristics?

Name _____
Date _____
Instructor _____

Fixed- and Voltage-Divider Bias of a BJT

OBJECTIVE

To determine the quiescent operating conditions of the fixed- and voltage-divider-bias BJT configurations.

EQUIPMENT REQUIRED

Instrument

DMM

Components

Resistors

(1) 680 Ω
(1) 2.7 kΩ
(1) 1.8 kΩ
(1) 6.8 kΩ
(1) 33 kΩ
(1) 1 MΩ

Transistors

(1) 2N3904 or equivalent
(1) 2N4401 or equivalent

Supplies

DC power supply

95

EQUIPMENT ISSUED

Item	Laboratory serial no.
DMM	
DC power supply	

RÉSUMÉ OF THEORY

Bipolar transistors operate in three modes: cutoff, saturation, and linear. In each of these modes, the physical characteristics of the transistor and the external circuit connected to it uniquely specify the operating point of the transistor. In the cutoff mode, there is only a small amount of reverse current from emitter to collector, making the device akin to an open switch. In the saturation mode, there is a maximum current flow from collector to emitter. The amount of that current is limited primarily by the external network connected to the transistor; its operation is analogous to that of a closed switch. Both of these operating modes are used in digital circuits.

For amplification with a minimum of distortion the linear region of the transistor characteristics is employed. A DC voltage is applied to the transistor, forward-biasing the base-emitter junction and reverse biasing the base-collector junction, typically establishing a quiescent point near or at the center of the linear region.

In this experiment, we will investigate two biasing networks: the fixed-bias and the voltage-divider-bias configuration. While the former is relatively simple, it has the serious drawback that the location of the Q-point is very sensitive to the forward current transfer ratio (β) of the transistor and temperature. Because there can be wide variations in beta and the temperature of the device or surrounding medium can change for a wide variety of reasons, it can be difficult to predict the exact location of the Q-point on the load line of a fixed-bias configuration.

The voltage-divider bias network employs a feedback arrangement that makes the base-emitter and collector-emitter voltages primarily dependent on the external circuit elements and not the beta of the transistor. Thus, even though the beta of individual transistors may vary considerably, the location of the Q-point on the load line will remain essentially fixed. The phrase "beta-independent biasing" is often used for such an arrangement.

PROCEDURE

Part 1. Determining β

a. Construct the network of Fig. 9.1 using the 2N3904 transistor. Insert the measured resistance values.

$R_{B(meas)}$ =_____
$R_{C(meas)}$ =_____

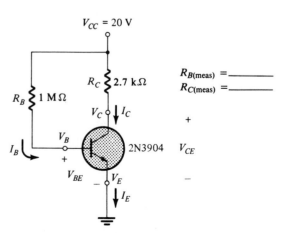

Figure 9-1

Procedure

b. Measure the voltages V_{BE} and V_{R_C}.

(measured) V_{BE} = _____

(measured) V_{R_C} = _____

c. Using the measured resistor values calculate the resulting base current using the equation

$$I_B = \frac{V_{R_B}}{R_B} = \frac{V_{CC} - V_{BE}}{R_B}$$

and the collector current using the equation

$$I_C = \frac{V_{R_C}}{R_C}$$

The voltage V_{R_B} was not measured directly for determining I_B because of the loading effects of the meter across the high resistance R_B.

Insert the resulting values of I_B and I_C in Table 9.1

d. Using the results of step 1(c) calculate the value of β and record in Table 9.1. This value of beta will be used for the 2N3904 transistor throughout this experiment.

Part 2. Fixed-Bias Configuration

a. Using the β determined in Part 1, calculate the currents I_B and I_C for the network of Fig. 9.1 using simply the measured resistor values, the supply voltage, and the above measured value for V_{BE}. That is, determine the theoretical values of I_B and I_C using simply the network parameters and the value of beta.

(calculated) I_B = _____

(calculated) I_C = _____

How do the calculated levels of I_B and I_C compare to those determined from measured voltage levels in Part 1(c)?

b. Using the results of step 2(a) calculate the levels of V_B, V_C, V_E, and V_{CE}.

(calculated) V_B = _____
(calculated) V_C = _____
(calculated) V_E = _____
(calculated) V_{CE} = _____

c. Energize the network of Fig. 9.1 and measure V_B, V_C, V_E, and V_{CE}.

(measured) V_B = _____
(measured) V_C = _____
(measured) V_E = _____
(measured) V_{CE} = _____

How do the measured values compare to the calculated levels of step 2(b)?

Record the measured value of V_{CE} in Table 9.1.

d. The next part of the experiment will essentially be a repeat of a number of the steps above for a transistor with a higher beta. Our goal is to show the effects of different beta levels on the resulting levels of the important quantities of the network. First the beta

Procedure

level for the other transistor, specifically a 2N4401 transistor, must be determined. Simply remove the 2N3904 transistor from Fig. 9.1 and insert the 2N4401 transistor, leaving all the resistors and voltage V_{CC} as in Part 1. Then measure the voltages V_{BE} and V_{R_C} and, using the same equations with measured resistor values, calculate the levels of I_B and I_C. Then determine the level of β for the 2N4401 transistor.

(measured) V_{BE} = _____
(measured) V_{R_C} = _____
(from measured) I_B = _____
(from measured) I_C = _____
(calculated) β = _____

Record the levels of I_B, I_C, and beta in Table 9.1. In addition measure the voltage V_{CE} and insert in Table 9.1.

TABLE 9.1

Transistor Type	V_{CE} volts	I_C mA	I_B μA	β
2N3904				
2N4401				

e. Using the following equations calculate the magnitude (ignore the sign) of the percent change in each quantity due to a change in transistors, specifically to one with a higher level of beta. Ideally, the important voltage and current levels should not change with a change in transistors but the fixed-bias configuration has a high sensitivity to changes in beta as will be reflected by the results. Place the results of your calculations in Table 9.2.

$$\% \Delta\beta = \frac{|\beta_{(4401)} - \beta_{(3904)}|}{|\beta_{(3904)}|} \times 100\% \qquad \% \Delta I_C = \frac{|I_{C(4401)} - I_{C(3904)}|}{|I_{C(3904)}|} \times 100\%$$

$$\% \Delta V_{CE} = \frac{|V_{CE(4401)} - V_{CE(3904)}|}{|V_{CE(3904)}|} \times 100\% \qquad \% \Delta I_B = \frac{|I_{B(4401)} - I_{B(3904)}|}{|I_{B(3904)}|} \times 100\%$$

(9.1)

TABLE 9.2
Percent Changes in β, I_C, V_{CE}, and I_B

%Δβ	%ΔI_C	%ΔV_{CE}	%ΔI_B

Part 3. Voltage-Divider Configuration

a. Construct the network of Fig. 9.2 using the 2N3904 transistor. Insert the measured value of each resistor.

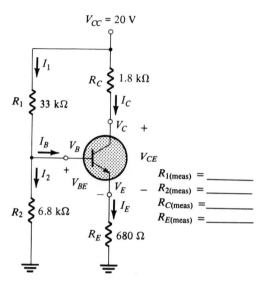

Figure 9-2

$R_{1(meas)} =$ _____
$R_{2(meas)} =$ _____
$R_{C(meas)} =$ _____
$R_{E(meas)} =$ _____

b. Using the beta determined in Part 1 for the 2N3904 transistor, calculate the theoretical levels of V_B, V_E, I_E, I_C, V_C, V_{CE}, and I_B for the network of Fig. 9.2 using an approach that will result in the highest level of accuracy for each quantity. Insert the results in Table 9.3.

TABLE 9.3

2N3904	V_B	V_E	V_C	V_{CE}	I_E (mA)	I_C (mA)	I_B (μA)
Calculated [Part 3(**b**)]							
Measured [Part 3(**c**)]							

c. Energize the network of Fig. 9.2 and measure V_B, V_E, V_C, and V_{CE} and record in Table 9.3. In addition, measure the voltages V_{R_1} and V_{R_2} to the highest degree of accuracy possible. That is, try to measure the quantities to the hundredth or thousandth place. Then calculate the currents I_E and I_C and the currents I_1 and I_2 (using $I_1 = V_{R_1}/R_1$ and $I_2 = V_{R_2}/R_2$) from the voltage readings and measured resistor values. Using the results for I_1 and I_2 calculate the current I_B using Kirchhoff's current law. Insert the calculated current levels for I_E, I_C, and I_B in Table 9.3.

In general, how do the calculated and measured values of Table 9.3 compare? Are there any significant differences that need to be explained?

d. Insert the measured value of V_{CE} and calculated values of I_C and I_B from step 3(**c**) in Table 9.4 along with the magnitude of beta from Part 1.

e. Replace the 2N3904 transistor of Fig. 9.2 with the 2N4401 transistor. Then measure the voltages V_{CE}, V_{R_C}, V_{R_1}, and V_{R_2}. Again, be sure to read V_{R_1} and V_{R_2} to the hundredth or thousandth place to insure an accurate determination of I_B. Then calculate I_C, I_1, I_2, and determine I_B. Complete Table 9.4 with the levels of V_{CE}, I_C, I_B, and beta for this transistor.

TABLE 9.4

Transistor Type	V_{CE} (volts)	I_C (mA)	I_B (μA)	β
2N3904				
2N4401				

f. Calculate the percent change in β, I_C, V_{CE}, and I_B from the data of Table 9.4. Use the formulas appearing in step 2(**e**), Eq. 9.1, and record your results in Table 9.5.

TABLE 9.5
Percent Changes in β, I_C, V_{CE}, and I_B

%Δβ	%ΔI_C	%ΔV_{CE}	%ΔI_B

Part 4. Computer Exercise

a. Perform a DC analysis of the network of Fig. 9.1 using PSpice, MicroCap II, or other appropriate software package. That is, find V_B, V_{CE}, I_B, and I_C using the 2N3904 transistor with the β determined in Part 1. The input file for Fig. 9.1 is the following for a PSpice analysis.

```
FIXED BIAS CONFIGURATION OF FIG. 9.1
VCC    3    0       20V
RB     3    1       1MEG
RC     3    2       2.7K
Q1     2    1       0        QN
.MODEL QN NPN (BF=(beta), IS=2E-15)
.DC VCC 20V 20V 1V
.PRINT DC V(1), V(2,0), I(RB), I(RC)
.OPTIONS NOPAGE
.END
```

b. Repeat the above analysis for the voltage-divider configuration of Fig. 9.2

Procedure

    c. How do the results of steps 4(**a**) and 4(**b**) (using the appropriate beta) compare with the measured values of the experiment?

Problems and Exercises

1. a. Compute the saturation current $I_{C_{sat}}$ for the fixed-bias configuration of Fig. 9.1.

(calculated) $I_{C_{sat}}$ = _____

    b. Compute the saturation current $I_{C_{sat}}$ for the voltage-divided bias configuration of Fig. 9.2.

(calculated) $I_{C_{sat}}$ = _____

    c. Are the saturation currents of parts (**a**) and (**b**) sensitive to the beta of the transistor or changes thereof?

2. For both the circuits investigated in this experiment, how did the Q point location (defined by I_C and V_{CE} on the collector characteristics) change when the 2N3904 transistor was replaced with the 2N4401? That is, how did the Q-point shift location when a transistor with a higher beta was substituted? In particular, did the Q-points move toward saturation (high I_C, low V_{CE}) or cut-off (low I_C, high V_{CE}) conditions?

3. **a.** Determine the ratio of the change in I_C, V_{CE}, and I_B due to changes in beta and complete Table 9.6. Use the results of Parts 2 and 3 to obtain the percent changes indicated.

TABLE 9.6

	$\dfrac{\%\Delta I_C}{\%\Delta\beta}$	$\dfrac{\%\Delta V_{CE}}{\%\Delta\beta}$	$\dfrac{\%\Delta I_B}{\%\Delta\beta}$
Fixed Bias			
Voltage-Divider			

b. One of the important goals of a good circuit design is to minimize the sensitivity of various circuit current and voltages to the beta variability of transistors. A figure of merit has been defined by the following equation which quantizes the percent change in collector current for a percent change in beta. In particular the smaller $S(\beta)$ the less the circuit will be affected by the change in beta.

$$S(\beta) = \frac{\%\Delta I_C}{\%\Delta\beta} \tag{9.2}$$

Referring to the results of Table 9.6, which network has the better stability factor $S(\beta)$? Is there a significant difference in level between the two stability factors?

c. Do the remaining sensitivities of Table 9.6 support the fact that one configuration is more stable than the other?

Problems and Exercises

4. **a.** For the fixed-bias configuration of Fig. 9.1 develop an equation for I_B in terms of the other elements (voltage source, resistors, β) of the network. Then establish an equation for I_C.

   b. Assuming I_1 and I_2 are much larger than I_B, permitting the approximation $I_1 \cong I_2$, develop an equation for I_C in terms of the other elements of the network of Fig. 9.2.

   c. Referring to the results of parts (a) and (b), is there an obvious reason why I_C is more sensitive to changes in beta in one configuration compared to the other?

Name _____
Date _____
Instructor _____

EXPERIMENT 10

Emitter and Collector Feedback Bias of BJTs

OBJECTIVE

To determine the quiescent operating conditions of the emitter and collector feedback bias BJT configurations.

EQUIPMENT REQUIRED

Instrument

DMM

Components

Resistors

(2) 2.2 kΩ
(1) 3 kΩ
(1) 390 kΩ
(1) 1 MΩ

Transistors

(1) 2N3904 or equivalent
(1) 2N4401 or equivalent

Supplies

DC power supply

EQUIPMENT ISSUED

Item	Laboratory serial no.
DMM	
DC power supply	

RÉSUMÉ OF THEORY

This experiment is an extension of Experiment 9. Two additional arrangements will be investigated in this experiment: the emitter bias and the collector feedback circuits.

Emitter Bias Circuit

The emitter bias configuration can be constructed using a single or a dual power supply. Both configurations offer increased stability over the fixed bias of Experiment 9. In particular, if the beta of the transistor times the resistance of the emitter resistor is large compared to the resistance of the base resistor, the emitter current becomes essentially independent of the beta of the transistor. Thus, if we exchange transistors in a properly designed emitter-bias circuit, the changes in I_C and V_{CE} should be small.

Collector Feedback Circuit

If we compare the collector feedback bias circuit configuration with the fixed bias of Experiment 9 it is noted that for the former, the base resistor is connected to the collector terminal of the transistor and not to the fixed supply voltage V_{CC}. Thus the voltage across the base resistance of the collector feedback configuration is a function of the collector voltage and in turn of the collector current. In particular, this circuit demonstrates the principle of negative feedback, in which a tendency of an output variable to increase or decrease will result in a reduction or increase in the input variable respectively. For instance, any tendency on the part of I_C to increase will reduce the level of V_C which in turn will result in a lower level of I_B offsetting the trend of I_C. The result is a design less sensitive to variations in its parameters.

PROCEDURE

Part 1. Determining β

a. Construct the network of Fig. 10.1 using the 2N3904 transistor. Insert the measured resistor values.

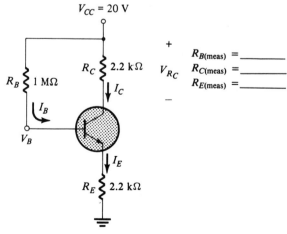

$R_{B(meas)}$ =_____
$R_{C(meas)}$ =_____
$R_{E(meas)}$ =_____

Figure 10-1

Procedure

b. Measure the voltages V_B and V_{R_C}.

(measured) V_B = _____

(measured) V_{R_C} = _____

c. Using the results of Part 1(b) and the measured resistor values calculate the resulting base currents I_B and I_C using the following equations:

$$I_B = \frac{V_{CC} - V_B}{R_B} \text{ and } I_C = \frac{V_{R_C}}{R_C}.$$

Record in Table 10.2.

(from measured) I_B = _____

(from measured) I_C = _____

d. Using the results of step 1(c) calculate the value of β and record in Table 10.2. This value of beta will be used for the 2N3904 transistor throughout the experiment.

(calculated) β = _____

Part 2. Emitter-Bias Configuration

a. Using the β determined in Part 1, calculate the values of I_B and I_C for the network of Fig. 10.1 using measured resistor values and the supply voltage V_{CC}. In order words, perform a theoretical analysis of the network. Insert the results in Table 10.1.

How do the theoretical values compare with the measured values of Part 1(c)?

b. Using the β determined in Part 1 calculate the levels of V_B, V_C, V_E, V_{BE}, and V_{CE} and insert in Table 10.1.

c. Next measure the voltages V_B, V_C, V_E, V_{BE}, and V_{CE} and insert in Table 10.2.

How do the calculated and measured results of Tables 10.1 and 10.2 compare for the 2N3904 transistor? In particular, comment on any results that do not compare well.

TABLE 10.1

Transistor Type	V_B volts	V_C volts	V_E volts	V_{BE} volts	V_{CE} volts	I_B μA	I_C mA
			Calculated Values				
2N3904							
2N4401							

TABLE 10.2

Transistor Type	V_B volts	V_C volts	V_E volts	V_{BE} volts	V_{CE} volts	I_B μA	I_C mA	β
	Measured Values					(Calc. from Measured Values)		
2N3904								
2N4401								

Procedure

d. Replace the 2N3904 transistor of Fig. 10.1 with the 2N4401 transistor and measure the resulting voltages V_B and V_{R_C}. Then calculate the currents I_B and I_C using measured resistance values and finally calculate the value of β for this transistor. This will be the value of beta used for the 2N4401 transistor throughout this experiment. Record the levels of I_B, I_C, and β in Table 10.2.

(measured) V_B = _____
(measured) V_{R_C} = _____

e. Using the beta determined in step 1(**d**), perform a theoretical analysis of Fig. 10.1 with the 2N4401 transistor. That is, calculate the levels of I_B, I_C, V_B, V_C, V_E, V_{BE}, and V_{CE} and insert in Table 10.1.

f. Energize the network of Fig. 10.1 with the 2N4401 transistor and measure V_B, V_C, V_E, V_{BE}, and V_{CE} and insert in Table 10.2.

How do the calculated and measured results of Tables 10.1 and 10.2 compare for the 2N4401 transistor? Discuss any results that appear different by more than 10%.

g. Calculate the percent change in β, I_C, V_{CE}, and I_B using the equations first presented in Experiment 9 and repeated here for convenience. Record the results in Table 10.3.

$$\% \Delta\beta = \frac{|\beta_{(4401)} - \beta_{(3904)}|}{|\beta_{(3904)}|} \times 100\% \qquad \% \Delta I_C = \frac{|I_{C(4401)} - I_{C(3904)}|}{|I_{C(3904)}|} \times 100\%$$

$$\% \Delta V_{CE} = \frac{|V_{CE(4401)} - V_{CE(3904)}|}{|V_{CE(3904)}|} \times 100\% \qquad \% \Delta I_B = \frac{|I_{B(4401)} - I_{B(3904)}|}{|I_{B(3904)}|} \times 100\%$$

TABLE 10.3
Percent Changes in β, I_C, V_{CE}, and I_B

%Δβ	%ΔI_C	%ΔV_{CE}	%ΔI_B

Part 3. Collector Feedback Configuration (R_E = 0 Ω)

a. Construct the network of Fig. 10.2 using the 2N3904 transistor. Insert the measured resistor values in Fig. 10.2.

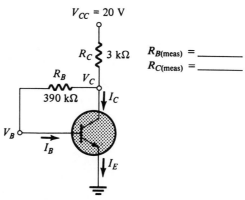

Figure 10-2 Collector Feedback

Procedure

b. Using the beta determined in Part 1, calculate the values of I_B, I_C, V_B, V_C, and V_{CE} and insert in Table 10.4.

c. Energize the network of Fig. 10.2, measure V_B, V_C, and V_{CE}, and insert in Table 10.5. Calculate the currents I_B and I_C using measured resistance values and the fact that $I_C \cong V_{R_C}/R_C$. Insert the current levels in Table 10.5.

How do the calculated and measured results of Tables 10.4 and 10.5 compare for the 2N3904 transistor?

d. Replace the 2N3904 transistor of Fig. 10.2 with the 2N4401 transistor of Part 1, calculate the values of I_B, I_C, V_B, V_C and V_{CE}, and insert in Table 10.4.

e. Energize the network of Fig. 10.2 with the 2N4401 transistor and measure V_B, V_C, and V_{CE}. Insert all measurements in Table 10.5. Calculate I_B and I_C from measured values and then insert the current levels in Table 10.5.

How do the calculated and measured results of Tables 10.4 and 10.5 compare for the 2N4401 transistor?

f. Calculate the percent changes in β, I_C, V_{CE}, and I_B using the equations of part 1(g). Record the results in Table 10.6.

TABLE 10.4

Transistor Type	Theoretical Calculated Values				
	V_B volts	V_C volts	V_{CE} volts	I_B μA	I_C mA
2N3904					
2N4401					

TABLE 10.5

Transistor Type	Measured Values			(Calc. from Measured Values)	
	V_B volts	V_C volts	V_{CE} volts	I_B μA	I_C mA
2N3904					
2N4401					

Procedure

TABLE 10.6
Percent Changes in β, I_C, V_{CE}, and I_B

%Δβ	%ΔI_C	%ΔV_{CE}	%ΔI_B

Part 4. Collector Feedback Configuration (with R_E)

a. Construct the network of Fig. 10.3 using the 2N3904 transistor. Insert the measured resistance values in Fig. 10.3.

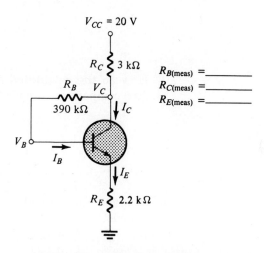

Figure 10-3 Collector feedback circuit.

b. Using the beta determined in Part 1, calculate the values of I_B, I_C, I_E, V_B, V_C, and V_{CE} and insert in Table 10.7.

c. Energize the network of Fig. 10.3, measure V_B, V_C, V_E, and V_{CE}, and insert in Table 10.8. In addition, calculate the currents I_B, I_C, and I_E from measured values using measured resistor values. Insert the current levels in Table 10.8.

How do the calculated and measured results of Tables 10.7 and 10.8 compare for the 2N3904 transistor?

d. Replace the 2N3904 transistor of Fig. 10.3 with the 2N4401 transistor and using the beta of Part 1 calculate the values of I_B, I_C, I_E, V_B, V_C and V_{CE} and insert in Table 10.7.

e. Energize the network of Fig. 10.3 with the 2N4401 transistor and measure V_B, V_C, V_E, and V_{CE} and insert in Table 10.8. In addition, calculate the currents I_B, I_C, and I_E from measured values using the measured resistor values. Insert the current levels in Table 10.8.

Procedure

How do the calculated and measured results of Tables 10.7 and 10.8 compare for the 2N4401 transistor?

f. Calculate the percent changes in β, I_C, V_{CE}, and I_B using the equations appearing in Part 1(**g**) and insert in Table 10.9.

TABLE 10.7

Transistor Type	Theoretical Calculated Values						
	V_B volts	V_C volts	V_E volts	V_{CE} volts	I_B µA	I_C mA	I_E mA
2N3904							
2N4401							

TABLE 10.8

Transistor Type	Measured Values				(Calc. from Measured Values)		
	V_B volts	V_C volts	V_E volts	V_{CE} volts	I_B µA	I_C mA	I_E mA
2N3904							
2N4401							

TABLE 10.9
Percent Changes in β, I_C, V_{CE}, and I_B

%Δβ	%ΔI_C	%ΔV_{CE}	%ΔI_B

Part 5. Computer Exercise

a. Perform a DC analysis of the network of Fig. 10.1 using PSpice, MicroCap II, or other appropriate software package. That is, find V_B, V_C, V_E, V_{CE}, I_B, and I_C using the 2N3904 transistor with the β determined in Part 1. The input file for Fig. 10.1 is the following with beta equal to the value determined in the experiment.

```
EMITTER BIAS CONFIGURATION OF FIG. 10.1
VCC   4    0    20V
RB    4    2    1MEG
RC    4    3    2.2K
RE    1    0    2.2K
Q1    3    2    1   QN
.MODEL QN NPN (BF=(beta), IS=2E-15)
.DC VCC 20V 20V 1V
.PRINT DC V(2), V(3), V(1), V(3,1),, I(RB), I(RC)
.OPTIONS NOPAGE
.END
```

b. Repeat the above analysis for the collector-feedback configuration of Fig. 10.3.

c. How do the results of steps 4(a) and 4(b) (using the appropriate beta) compare with the measured values of the experiment?

Problems and Exercises

1. a. Compute the saturation current $I_{C_{sat}}$ for the emitter-bias configuration of Fig. 10.1.

   (calculated) $I_{C_{sat}}$ = _____

   b. Compute the saturation current $I_{C_{sat}}$ for the collector-feedback configuration of Fig. 10.2.

   (calculated) $I_{C_{sat}}$ = _____

   c. Compute the saturation current $I_{C_{sat}}$ for the collector-feedback configuration of Fig. 10.3.

   (calculated) $I_{C_{sat}}$ = _____

   d. What is the effect of beta on the calculations of **a**, **b**, and **c** of this exercise?

2. For the three configurations investigated in this experiment, how did the Q-point location (defined by I_C and V_{CE}) change when the 2N3904 transistor was replaced with the 2N4401? That is, how did the Q-point shift position when a transistor with a higher beta was substituted? In particular, did the Q-points move toward saturation (high I_C, low V_{CE}) or cut-off (low I_C, high V_{CE}) conditions?

Problems and Exercises

3. a. Determine the ratio of the change in I_C, V_{CE}, and I_B due to changes in beta and complete Table 10.10. Use the results of Parts 2, 3, and 4 to obtain the percent changes indicated.

TABLE 10.10

	$\dfrac{\%\Delta I_C}{\%\Delta \beta}$	$\dfrac{\%\Delta V_{CE}}{\%\Delta \beta}$	$\dfrac{\%\Delta I_B}{\%\Delta \beta}$
Emitter bias			
Collector feedback ($R_E = 0\ \Omega$)			
Collectector feedback (with R_E)			

b. How does the figure of merit defined by Eq. 9.2 (repeated here for convenience) compare for each configuration of Table 10.10?

$$S(\beta) = \frac{\%\Delta I_C}{\%\Delta \beta}$$

Which appears to have the better stability factor?

c. Do the remaining sensitivities [$S(\beta)$] of Table 10.10 support the fact that one configuration is more stable than the other?

4. a. For the emitter-bias configuration of Fig. 10.1 develop an equation for I_C in terms of the other elements (resistors, V_{CC}, β) of the network. Use the fact that $(\beta + 1) \cong \beta$.

b. Divide the numerator and denominator of the equation obtained in **a** by β.

c. Based on the results of **b** what relationship must exist between the elements of the network to minimize the effect of changing levels of β on the level of I_C?

5. a. For the collector feedback configuration of Fig. 10.2 develop an equation for I_C in terms of the other elements (resistors, V_{CC}, β) of the network. Use the fact that $(\beta + 1) \cong \beta$.

Procedure

    b. Divide the numerator and denominator of the equation obtained in part **a** by β.

    c. Based on the results of **b** what relationship must exist between the elements of the network to minimize the effect of changing levels of β on the level of I_C?

6. a. For the collector feedback configuration of Fig. 10.3 develop an equation for I_C in terms of the other elements (resistors, V_{CC}, β) of the network. Use the fact that $(\beta + 1) \cong \beta$.

    b. Divide the numerator and denominator of the equation obtained in part **a** by β.

    c. Based on the results of **b** what relationship must exist between the elements of the network to minimize the effect of changing levels of β on the level of I_C?

7. Comparing the results of Exercises 4(c), 5(c), and 6(c) which configuration would appear to have the least sensitivity to changes in beta for resistor values of about the same magnitude?

Does the above conclusion compare favorably with the conclusion of Exercise 3(b)?

Name _____
Date _____
Instructor _____

EXPERIMENT 11

Design of BJT Bias Circuits

OBJECTIVE

To design a Collector-Feedback, Emitter-Bias, and Voltage-Divider-Bias BJT transistor network.

EQUIPMENT REQUIRED

Instruments

(1) DMM

Components

Resistors

Since this is a design experiment a number of the required resistors are not specified in the equipment list. They will have to be requested from the stockroom once their values are determined.

(1) 300 Ω, 1.2 kΩ, 1.5 kΩ, 3 kΩ, 15 kΩ, 100 kΩ
(1) 1 MΩ potentiometer
Other resistors as required by the designs

Transistors

(1) 2N3904 or equivalent
(1) 2N4401 or equivalent

Supplies

(1) DC Power Supply

EQUIPMENT ISSUED

Item	Laboratory serial no.
DMM	
DC power supply	

RÉSUMÉ OF THEORY

In this experiment we will make a preliminary design of a collector-feedback, emitter-bias, and voltage-divider bias BJT transistor configuration. Unlike analysis, where the circuit is given and the response of the circuit variables is asked for, in circuit design, the desired circuit responses are specified and a circuit that yields the desired variables is to be constructed.

The approach taken in this experiment will show that circuit design is often a series of compromises. The most stable network may not result in an acceptable level of ac gain. The resistor values that the theoretical calculations suggest would be the best for a particular design may not be commercially available. One set of resistor values may result in the most stable system with excellent gain characteristics but result in a low conversion efficiency as defined by $\eta\% = P_o(ac)/P_i(dc) \times 100\%$. The designer must be aware of the consequences of making a certain choice and which characteristics of the design are the most vital for the particular application.

For a specified Q-point, β, and an appropriate level for V_E, the following equations can be applied as a DC design sequence:

Collector Feedback:

$$R_C = \frac{V_{CC} - V_{CE_Q}}{I_{C_Q}} \tag{11.1}$$

$$R_B = \frac{V_{R_B}}{I_B} = \frac{V_{CE_Q} - V_{BE}}{\frac{I_{C_Q}}{\beta}} = \beta \left[\frac{V_{CE_Q} - V_{BE}}{I_{C_Q}} \right] \tag{11.2}$$

Emitter-Bias:

$$R_E = \frac{V_E}{I_{C_Q}} \tag{11.3}$$

$$V_C = V_{CE_Q} + V_E \tag{11.4}$$

$$R_C = \frac{V_{R_C}}{I_{C_Q}} = \frac{V_{CC} - V_C}{I_{C_Q}} \tag{11.5}$$

$$R_B = \frac{V_{R_B}}{I_B} = \frac{V_{CC} - V_{BE} - V_E}{\frac{I_{C_Q}}{\beta}} = \beta \left[\frac{V_{CC} - V_{BE} - V_E}{I_{C_Q}} \right] \tag{11.6}$$

Voltage-Divider Bias:

$$R_C = \frac{V_{CC} - V_{CE_Q} - V_E}{I_{C_Q}} \qquad (11.7)$$

$$R_E = \frac{V_E}{I_{C_Q}} \qquad (11.8)$$

Assuming $\beta R_E > 10 R_2$ will result in

$$V_B = \frac{R_2 V_{CC}}{R_1 + R_2} = V_{BE} + V_E \qquad (11.9)$$

The magnitude of R_1 and R_2 is defined by a chosen relationship between the two.

Design Criteria

For each of the above configurations the following defines the relative stability of the system:

Collector-Feedback:

$$\text{Increasing } \frac{R_B}{\beta R_C} \text{ decreases stability} \qquad (11.10)$$

Emitter-Bias:

$$\text{Increasing } \frac{R_B}{\beta R_E} \text{ decreases stability} \qquad (11.11)$$

Voltage-Divider Bias:

$$\text{Increasing } \frac{R_1 \| R_2}{\beta R_E} \text{ decreases stability} \qquad (11.12)$$

PROCEDURE

Part 1. Collector-Feedback Configuration

Circuit Specifications:

$$V_{CC} = 15 \text{ V}$$
$$I_{C_Q} = 5 \text{ mA}$$
$$V_{CE_Q} = 7.5 \text{ V}$$

Design Procedure:

a. From the given specifications, determine the required value of R_C for the collector feedback network of Fig. 11.1.

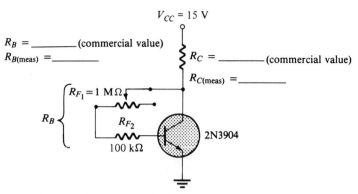

Figure 11-1

Determine the closest commercial value (available in the laboratory) and record below and in Fig. 11.1. Obtain the chosen resistor and insert the measured resistance value in the space provided in Fig. 11.1.

(calculated) $R_C =$ _____

(commercial value) $R_C =$ _____

b. Connect a 100 kΩ resistor and the 1 MΩ potentiometer, set to a maximum, in series, as in Fig. 11.1. Use the commercial value of Part 1 for R_C. With the power on, adjust the potentiometer until $V_{CE} = 7.5$ V using the 2N3904 transistor.

c. Turn off the supply and disconnect the 100 kΩ resistor from the transistor base connection and measure the combined resistance of R_{F_1} and R_{F_2}. Select a commercial resistor close to this combined level (that is available in the laboratory) and record its nominal value as R_B. In addition include its measured value on Fig. 11.1.

(measured) $R_B = R_{F_1} + R_{F_2} =$ _____

(commercial value) $R_B^2 =$ _____

d. Replace R_{F_1} and R_{F_2} in the assembled network with the commercial R_B value selected in part 1(**c**). Then make the measurements and calculations listed below. Use measured values for the

Procedure

resistance levels. Determine I_{C_Q} from $I_{C_Q} = V_{R_C}/R_C$ and I_B from $I_B = (V_{CE} - V_{BE})/R_B$.

(measured) V_{R_C} = _____

(measured) V_{CE_Q} = _____

(from measured) I_{C_Q} = _____

(calculated) β = _____

e. Referring to the results of Part 1(**d**), how do the resulting values of I_{C_Q} and V_{CE_Q} compare to the specified values?

Would you consider the design to be a relatively good one?

f. The Résumé of Theory introduced the ratio $\dfrac{R_B}{\beta R_C}$ as an indication of the relative stability of the system. Determine the ratio and insert below. It will be examined in a later part of the experiment.

(calculated) $R_B/\beta R_C$ _____

g. Rebuild Fig. 11.1 with the 2N4401 transistor and the same value of R_C and repeat Parts 1(**b**) and 1(**c**).

(measured) $R_{F_1} + R_{F_2}$ = _____

(commercial value) R_B = _____

h. Referring to Part 1(**g**), did the value of R_B change with the use of transistor with a higher β?

Why did the value of R_C remain the same for both transistors?

i. Repeat Part 1(d) using the 2N4401 transistor.

(measured) V_{R_C} = _____

(measured) V_{CE_Q} = _____

(from measured) I_{C_Q} = _____

(calculated) β = _____

j. Referring to the results of Part 1(i) how do the resulting values of I_{C_Q} and V_{CE_Q} compare to the specified values?

How would you compare this design to the design using the 2N3904 transistor?

k. Determine the ratio $\dfrac{R_B}{\beta R_C}$ for this configuration and compare to the level calculated for the 2N3904 transistor.

(calculated) $\dfrac{R_B}{\beta R_C}$ (2N4401) = _____

(calculated) $\dfrac{R_B}{\beta R_C}$ (2N3904) = _____

What is interesting about the results just obtained? What do the results suggest about the two networks when we discuss their relative stability?

l. Rebuild the network of Fig. 11.1 with the level of R_C and R_B calculated for the 2N3904 transistor. However, this time insert the 2N4401 transistor so we can measure the change in I_C due to a change in beta. Energize the network and measure the voltage V_{R_C}. Using the measured resistance value for R_C calculate I_{C_Q} then determine $S(\beta) = \dfrac{\%\Delta I_C}{\%\Delta \beta}$ for the first design.

(calculated) $S(\beta) =$ _____

Part 2. Emitter-Bias Configuration

Circuit Specifications:

$$V_{CC} = 15 \text{ V}$$
$$I_{C_Q} = 5 \text{ mA}$$
$$V_{CE_Q} = 7.5 \text{ V}$$

Design Procedure

In this case we will employ a common design rule that $V_E = 0.1\ V_{CC}$. Note, in particular, that the Q-point location on the load line is the same as defined for the collector-feedback configuration.

  a. For the given specifications, calculate the required value of R_C for the emitter-biased configuration of Fig. 11.2. Determine the closest commercial value (available in the laboratory) and record below and in Fig. 11.2. Obtain the resistor and insert its measured value in the space provided in Fig. 11.2.

(calculated) $R_C =$ _____
(commercial value) $R_C =$ _____

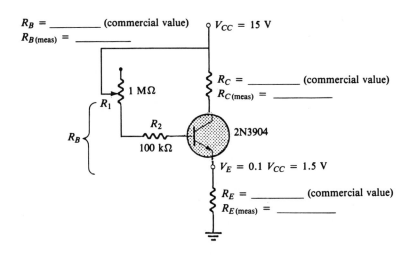

Figure 11-2

b. Using $V_E = 0.1\, V_{CC} = 1.5$ V, calculate the required value of R_E for the network of Fig. 11.2. Determine the closest commercial value (available in the laboratory) and record below and in Fig. 11.2. Obtain the chosen resistor and insert the measured value in the space provided in Fig. 11.2.

(calculated) $R_E = $ _____

(commercial value) $R_E = $ _____

c. Connect a 100 kΩ resistor and the 1 MΩ potentiometer (set to a maximum) in series, as in Fig. 11.2. With the power on, adjust the potentiometer until $V_{CE} = 7.5$ V using the 2N3904 transistor. Use the commercial values of R_C and R_E as determined in steps 2(a) and 2(b).

d. Turn off the supply and disconnect the R_2 from the base of the transistor. Measure the series resistance determined by $R_1 + R_2$ and record below. Then determine the closest available commercial value (available in the laboratory) and insert below and on Fig. 11.2. Obtain the chosen resistor and insert the measured value in Fig. 11.2.

(measured) $R_B = R_1 + R_2 = $ _____

(commercial value) $R_B = $ _____

e. Rebuild the network of Fig. 11.2 with all the resulting commercial values determined in the above steps. Measure the voltages V_{R_C} and V_{CE} and calculate the current I_C using the measured resis-

Procedure

tance value for R_C. In addition, measure the voltage V_B and calculate the current I_B. Finally calculate the beta of the transistor.

(measured) V_{R_C} = _____
(measured) V_{CE} = _____
(from measured) I_C = _____
(from measured) I_B = _____
(calculated) β = _____

f. Referring to the results of Part 2(**e**) how do the resulting values of I_{C_Q} and V_{CE_Q} compare to the specified values?

Would you consider the design to be a relatively good one?

g. The Résumé of Theory introduced the ratio $R_B/\beta R_E$ as an indication of the relative stability of the system. Determine the ratio and insert below.

(calculated) $R_B/\beta R_E$ = _____

h. Rebuild Fig. 11.2 with the 2N4401 transistor and the same value of R_C and R_E and repeat Parts 2(**c**) and 2(**d**).

(calculated) $R_B =$ _____

(commercial value) $R_B =$ _____

i. Referring to Part 1(h), did the value of R_B change with the use of transistor with a higher beta?

    Why did the value of R_C and R_E remain the same for both transistors?

j. Repeat Part 2(e) using the 2N4401 transistor.

(measured) $V_{R_C} =$ _____

(measured) $V_{CE_Q} =$ _____

(calculated) $I_{C_Q} =$ _____

(calculated) $\beta =$ _____

k. Referring to the results of Part 2(j) how do the resulting values of I_{C_Q} and V_{CE_Q} compare to the specified values?

l. Since the Q-point for this design is the same for the collector-feedback circuit of Part 1, the magnitudes of beta for each transistor should also be very close for each configuration. Is this conclusion verified by the results of Parts 2(e) and 2(j)?

Procedure

m. Determine the ratio $R_B/\beta R_E$ for this configuration and compare to the level calculated for the 2N3904 transistor.

(calculated) $R_B/\beta R_E$ (2N4401) = _____

(calculated) $R_B/\beta R_E$ (2N3904) = _____

What is interesting about the results just obtained? What do the results suggest about the two networks when we discuss their relative stability?

n. Rebuild the network of Fig. 11.2 with the level of R_C, R_E, and R_B calculated for the 2N3904 transistor. However, this time insert the 2N4401 transistor so we can measure the change in I_C due to a change in beta. Energize the network and measure the voltage V_{R_C}. Using the measured resistance value for R_C calculate I_{C_Q} and then determine $S(\beta) = \dfrac{\%\Delta I_C}{\%\Delta \beta}$ for the first design.

(calculated) $S(\beta)$ = _____

Part 3. Voltage-Divider Configuration

Circuit Specifications:

$$V_{CC} = 15 \text{ V}$$
$$I_C = 5 \text{ mA}$$
$$V_{CE} = 7.5 \text{ V}$$

Design Procedure

Again, the Q-point is the same as defined in Parts 1 and 2. In addition, we will continue to apply the general rule that $V_E = 0.1\, V_{CC}$.

a. For the given specifications, calculate the required value of R_C for the voltage-divider configuration of Fig. 11.3. Determine the closest commercial value (available in the laboratory) and record below and in Fig. 11.3. Insert the measured resistor value in the space provided in Fig. 11.3.

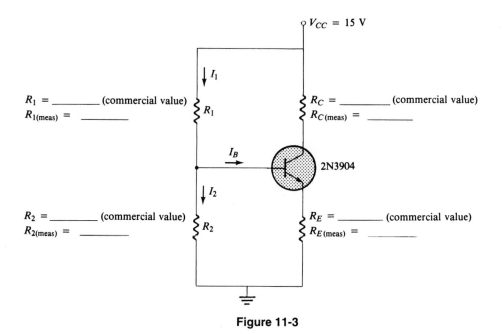

Figure 11-3

(calculated) $R_C =$ _____

(commercial value) $R_C =$ _____

b. Using $V_E = 0.1\, V_{CC} = 1.5$ V, calculate the required value of R_E for the network of Fig. 11.3. Determine the closest commercial value (available in the laboratory) and record below and in Fig. 11.3. Insert the measured resistor value in the space provided in Fig. 11.3.

Procedure

(calculated) $R_E =$ _____

(commercial value) $R_E =$ _____

c. Assume $\beta R_E > 10R_2$ permits the use of Eq. 11.9 of the Résumé of Theory to define a relationship between R_1 and R_2. Determine that relationship using the circuit specifications.

Relationship: _____

d. Using $\beta = 100$ and R_E as determined in Part 3(b) (commercial value) calculate the maximum value of R_2 to satisfy the relationship $\beta R_E > 10R_2$.

(calculated) $R_2 =$ _____

Choose the nearest standard commercial value (available in the laboratory) for R_2 and calculate the required value of R_1 using the commercial value for R_2.

(commercial value) $R_2 =$ _____

(calculated) $R_1 =$ _____

Choose the next lowest standard commercial value (available in the laboratory) for R_1 and list below:

(commercial value) $R_1 =$ _____

e. Using the chosen standard commercial values construct the network of Fig. 11.3 and insert the measured value for each resistor. Measure the voltages V_{R_C} and V_{CE_Q} and calculate the current I_{C_Q} using the measured resistance value for R_C. Finally, make a careful measurement of V_{R_1} and V_{R_2} (at least to hundredths place) and calculate the currents I_1 and I_2 to the same level of accuracy using measured resistor values. Then determine I_B and calculate β.

(measured) V_{R_C} = _____

(measured) V_{CE_Q} = _____

(calculated) I_{C_Q} = _____

(calculated) β = _____

Are the resulting values of I_{C_Q} and V_{CE_Q} relatively close to the specified values?

If not, try to explain why. How would you correct the situation?

f. The Résumé of Theory introduced the ratio $R_1||R_2/\beta R_E$ as an indication of the relative stability of the design. Determine the ratio and insert below.

(calculated) $R_1||R_2/\beta R_E$ = _____

g. Rebuild Fig. 11.3 with the 2N4401 transistor and repeat Part 3(**e**).

Problems and Exercises

$\qquad$ (measured) $V_{R_C} =$ _____

$\qquad$ (measured) $V_{CE_Q} =$ _____

$\qquad$ (calculated) $I_{C_Q} =$ _____

$\qquad$ (calculated) $\beta =$ _____

h. Referring to the results of Part 3(**g**), how do the resulting values of I_{C_Q} and V_{CE_Q} compare to the specified values even though β has been significantly increased?

i. How do the levels of beta for this part of the experiment compare to the levels determined in Parts 1 and 2?

j. Determine the ratio for this configuration and compare to the level calculated for the 2N3904 transistor.

$\qquad$ (calculated) $R_1 || R_2 / \beta R_E$ (2N4401) = _____

$\qquad$ (calculated) $R_1 || R_2 / \beta R_E$ (2N3904) = _____

What do the results suggest about the impact of β on the stability of the voltage-divider configuration?

k. Determine the stability factor $S(\beta) = \dfrac{\%\Delta I_C}{\%\Delta \beta}$ using the data of parts 3(**e**) and 3(**g**).

$\qquad$ (calculated) $S(\beta) =$ _____

Problems and Exercises

1. a. For each configuration designed in this experiment insert the resulting measured values of I_{C_Q} and V_{CE_Q} in Table 11.1.

TABLE 11.1

Configuration	I_{C_Q}	V_{CE_Q}
Collector-feedback-bias		
Emitter-bias		
Voltage-divider-bias		

Based on the above as compared to the specified values of I_{C_Q} = 5 mA and V_{CE_Q} = 7.5 V do you feel satisfied with your design effort?

Which design came closest to the specified levels?

2. Using the results of the experiment, complete Table 11.2 for the 2N4401 Transistor.

TABLE 11.2

Configuration		Stability factor
Collector feedback	$R_B/\beta R_C =$	$S(\beta) =$
Emitter bias	$R_B/\beta R_E =$	$S(\beta) =$
Voltage-divider bias	$R_1||R_2/\beta R_E =$	$S(\beta) =$

Is there a consistency between the stability factors of the two columns—that is, if relatively small in one column, is it relatively small in the other?

Which configuration demonstrated through $S(\beta)$ that it was the least sensitive to changes in beta? Was this expected? Why?

Which configuration has the least stability? Was this expected? Why?

The next three problems will derive the stability criteria appearing in the Résumé of Theory.

3. For the collector-feedback configuration derive an equation for the current I_C in terms of the network parameters. Then explain why the smaller the ratio $R_B/\beta R_C$ is, the less the sensitivity of I_C is to changes in beta.

4. For the emitter-bias configuration derive an equation for the current I_C in terms of the network parameters. Then explain why the smaller the ratio $R_B/\beta R_E$ is, the less the sensitivity of I_C is to changes in beta.

5. Defining the base voltage as V_B write an equation for I_1 and I_2 of the voltage-divider configuration in terms of the network parameters. Use the resulting equations to write an equation for I_B. Substitute the fact

that $V_B \cong V_{BE} + I_C R_E$ and rewrite the equation for I_B in a form that will support the fact that the smaller the ratio $R_1 \| R_2 / \beta R_E$ is, the less sensitive the current I_C is to changes in beta. Use the notation:

$$R_1 \| R_2 = \frac{1}{\frac{1}{R_1} + \frac{1}{R_2}}$$

Name _____
Date _____
Instructor _____

EXPERIMENT 12

JFET Characteristics

OBJECTIVE

To obtain the output and transfer characteristics for a JFET transistor.

EQUIPMENT REQUIRED

Instruments

DMM
Curve tracer (if available)

Components

Resistors

(1) 100-Ω
(1) 1-kΩ
(1) 10-kΩ
(1) 5 kΩ potentiometer
(1) 1 MΩ potentiometer

Transistor

(1) 2N4416 (or equivalent)

Supplies

DC power supply
9V battery with snap-on leads

Exp. 12 / JFET Characteristics

EQUIPMENT ISSUED

Item	Laboratory serial no.
DMM	
DC power supply	

RESUMÉ OF THEORY

The junction field-effect transistor (JFET) is a unipolar conduction device. The current carriers are either electrons in an n-channel JFET or holes in a p-channel JFET. In the n-channel JFET the conduction path is an n-doped material, germanium or silicon, while in the p-channel the conduction path is p-doped germanium or silicon. Conduction through the channel is controlled by the depletion region established by oppositely doped regions in the channel. The channel is connected to two terminals, referred to as the drain and the source, respectively. For n-channel JFETs, the drain is connected to a positive voltage, and the source to a negative voltage, to establish a flow of conventional current in the channel. The polarities of the applied voltages for the p-channel JFET are opposite to those of the n-channel JFET.

A third terminal, referred to as the gate terminal, provides a mechanism for controlling the depletion region and thereby the width of the channel through which conventional flow can exist between the drain and source terminals. For an n-channel JFET, the more negative the gate-to-source voltage is, the smaller the channel width is and the less the drain-do-source current is.

This experiment will establish the relationships between the various voltages and currents flowing in a JFET. The nature of these relationships determines the range of JFET applications.

PROCEDURE

Part 1. Measurement of the Saturation Current I_{DSS} and Pinch-Off Voltage V_P.

a. Construct the network of Fig. 12.1. Insert the measured value of R. The 10 kΩ resistor in the input circuit is included to protect the gate circuit if the 9 V battery is applied with the wrong polarity and the potentiometer is set on its maximum value.

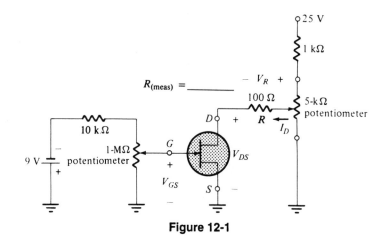

Figure 12-1

Procedure

b. Vary the 1 MΩ potentiometer until $V_{GS} = 0$ V. Recall that $I_D = I_{DSS}$ when $V_{GS} = 0$ V.

c. Set V_{DS} to 8 V by varying the 5 kΩ potentiometer. Measure the voltage V_R.

(measured) $V_R = $ _____

d. Calculate the saturation current from $I_{DSS} = I_D = V_R/R$ using the measured resistor value and record below

(from measured) $I_{DSS} = $ _____

e. Maintain V_{DS} at about 8 V and reduce V_{GS} until V_R drops to 1 mV. At this level $I_D = V_R/R = 1$ mV/100 Ω $= 10$ μA $\cong 0$ mA compared to typical operating levels. Recall that V_P is the voltage V_{GS} that results in $I_D = 0$ mA. Record the pinch-off voltage below:

(measured) $V_P = $ _____

f. Check with two other groups in your laboratory area and record their levels of I_{DSS} and V_P.

1. $I_{DSS} = $ _____ , $V_P = $ _____
2. $I_{DSS} = $ _____ , $V_P = $ _____

Based on the above, are I_{DSS} and V_P the same for all 2N4416 transistors?

g. Using the determined values of I_{DSS} and V_P, sketch the transfer characteristics for the device in Fig. 12.2 using Shockley's equation. Plot at least 5 points on the curve.

$$I_D = I_{DSS}\left(1 - \frac{V_{GS}}{V_P}\right)^2 \qquad (12.1)$$

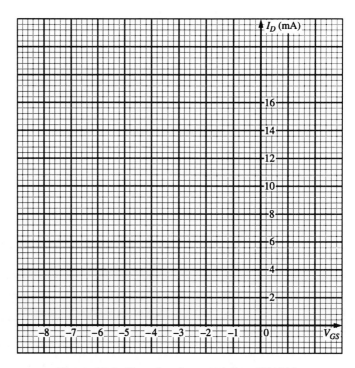

Figure 12-2 Transfer characteristics: 2N4416

Part 2. Output Characteristics

This part of the experiment will determine the I_D versus V_{DS} characteristics for an n-channel JFET.

    a. Using the network of Fig. 12.1, vary the two potentiometers until $V_{GS} = 0$ V and $V_{DS} = 0$ V. Determine I_D from $I_D = V_R/R$ using the measured value of R and record in Table 12.1.

    b. Maintain V_{GS} at 0 V and increase V_{DS} through 14 V (in 1 volt steps) and record the calculated value of I_D. Be sure to use the measured value of the 100-Ω resistance in your calculations.

    c. Vary the 1-MΩ potentiometer until $V_{GS} = -1$ V. Maintaining V_{GS} at this level, vary V_{DS} through the levels of Table 12.1 and record the calculated value of I_D.

    d. Repeat step **c** for the values of V_{GS} appearing in Table 12.1. Discontinue the process once V_{GS} exceeds V_P.

    e. Plot the output characteristics for the JFET on the graph of Fig. 12.3.

    f. Does the plot verify the conclusions of Part 1? That is, is the average value of I_D for $V_{GS} = 0$ V relatively close to I_{DSS}? Is the value of V_{GS} that results in $I_D = 0$ mA close to V_P?

Procedure

TABLE 12.1

V_{GS} (V)	0	−1.0	−2.0	−3.0	−4.0	−5.0	−6.0
V_{DS} (V)	I_D (mA)	I_D (mA)	I_D (mA)	I_D (mA)	I_D (mA)	I_D (mA)	I_D (mA)
0.0							
1.0							
2.0							
3.0							
4.0							
5.0							
6.0							
7.0							
8.0							
9.0							
10.0							
11.0							
12.0							
13.0							
14.0							

I_{DSS} (Fig. 12.3) = _____
I_{DSS} (Part 1) = _____
V_P (Fig. 12.3) = _____
V_P (Part 1) = _____

Part 3. Transfer Characteristics

This part of the experiment will determine the I_D vs. V_{GS} transfer characteristics frequently used in the analysis of JFET networks. Ideally, the transfer characteristics as determined by Shockley's equation assume that the effect of V_{DS} can be ignored and the characteristic curves of Fig. 11.3 for a given V_{GS} are considered horizontal. The following will show that the transfer curve does vary slightly with V_{DS} but not to the point where concern should develop about using Shockley's equation.

For this part of the experiment all the data can be obtained from Table 12.1. There is no experimental work in this part.

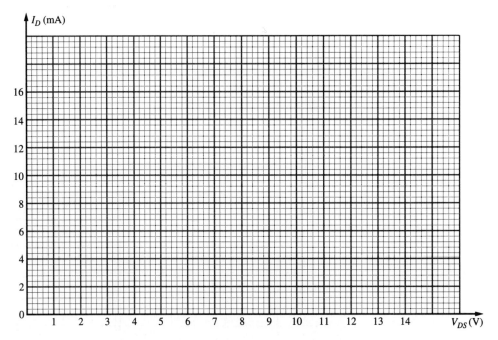

Figure 12-3 Drain-current curve: 2N4416.

a. At $V_{DS} = 3$ V record the values of I_D for the range of V_{GS} in Table 12.2 using the data of Table 12.1.

TABLE 12.2

V_{DS}	3 V	6 V	9 V	12 V
V_{GS}	I_D(mA)	I_D(mA)	I_D(mA)	I_D(mA)
0 V				
–1 V				
–2 V				
–3 V				
–4 V				
–5 V				
–6 V				

b. Repeat Part **a** for $V_{DS} = 6$ V, 9 V, and 12 V.
c. For each level of V_{DS} plot I_D vs. V_{GS} on the graph of Fig. 12.4. Plot each curve carefully and label each curve with the value of V_{DS}.
d. Is it reasonable (on an approximate basis) to assume the family of curves of Fig. 12.4 can be replaced by a single curve defined by Shockley's equation?

Procedure

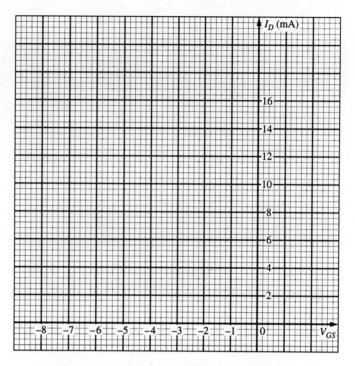

Figure 12-4 Pinch-off voltage curve: 2N4416

Part 4. Determination of the JFET Characteristics Using a Commercial Curve Tracer

a. If available, use the curve tracer to obtain an output set (I_D vs. V_{DS}) of characteristics for the 2N4416 JFET.

b. Reproduce the characteristics on the graph of Fig. 12.5.

c. Compare the characteristics to those obtained in Part 2, Figure 12.3. Note that the scales are the same to permit a direct comparison.

d. From your data obtained in Figure 12.5 draw the transfer characteristics in Figure 12.6. Compare this graph with Figure 12.4 in Part 3. Use as many data points from Fig. 12.5 as you feel are required to obtain the desired curves.

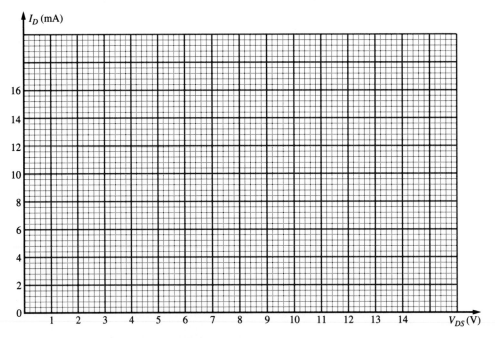

Figure 12-5 Drain-source characteristic: 2N4416.

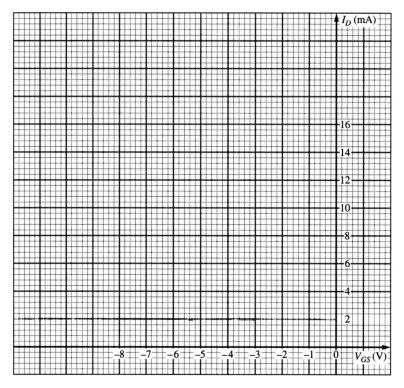

Figure 12-6 Transfer characteristic: 2N4416.

Problems and Exercises

1. Given I_D and V_{GS} at a particular point on Shockley's curve can the values of I_{DSS} and V_P be determined? If so, how? If not, why not?

2. a. Write Shockley's equation in a form that will provide V_{GS} in terms of I_{DSS}, V_P, and I_D.

   b. Given $I_{DSS} = 10$ mA, $V_P = -5$ V, and $I_D = 4$ mA, find the value of V_{GS}.

   (calculated) $V_{GS} =$ _____

3. The transconductance, g_m, of a JFET is an important quantity in the ac analysis of JFET amplifiers. Its magnitude is defined by the slope of Shockley's equation at a point on the characteristics. The application of calculus techniques to Shockley's equation will result in the following equation for g_m:

$$g_m = g_{mo}(1 - V_{GS}/V_P) \qquad (12.1)$$

with

$$g_{mo} = \frac{2 I_{DSS}}{|V_P|} \qquad (12.2)$$

which is the transconductance at $V_{GS} = 0$ V.

   a. Using the experimental results of Part 1 determine g_{mo}.

   (calculated) $g_{mo} =$ _____

b. Referring to the transfer curve of Fig. 12.2, is the slope of Shockley's equation a maximum at $V_{GS} = 0$ V?

Base on the above, can we assume that g_{m_o} calculated in part **a** of this exercise is the maximum value of g_m?

c. Determine g_m at $V_{GS} = V_P$.

(calculated) $g_m =$ _____

Referring to the transfer curve of Fig. 12.2, is the slope of Shockley's equation a minimum at $V_{GS} = V_P$? In fact, what slope would you expect it to have exactly at $V_{GS} = V_P$?

d. Determine g_m at $V_{GS} = 1/4\ V_P$, $1/2\ V_P$, and $3/4\ V_P$, and plot the curve of g_m on Fig. 12.7.

Problems and Exercises

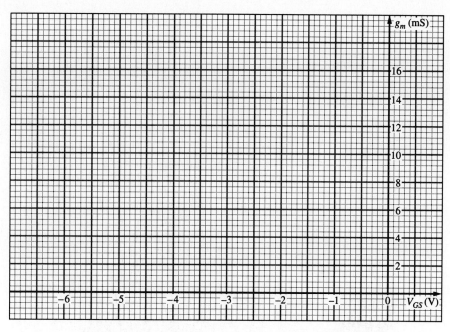

Figure 12-7 Transconductance versus V_{GS} of 2N4416.

e. Referring to the transfer curve of Fig. 12.2, does the slope increase with less negative values of V_{GS}? Is your conclusion verified by the plot of Fig. 12.7?

f. What does the fact that the graph of Fig. 12.7 is a straight line tell you about the curve resulting from Shockley's equation?

Name _____
Date _____
Instructor _____

EXPERIMENT 13

JFET Bias Circuits

OBJECTIVE

To analyze the fixed-, self-, and voltage-divider-bias JFET networks.

EQUIPMENT REQUIRED

Instrument

DMM

Components

Resistors

(1) 1-kΩ
(1) 1.2-kΩ
(1) 2.2-kΩ
(1) 3-kΩ
(1) 10-kΩ
(1) 10-MΩ
(1) 1 kΩ potentiometer

Transistors

(1) JFET 2N4166 (or equivalent)

Supplies

DC power supply
9 V battery with snap-on leads

153

Exp. 13 / JFET Bias Circuits

EQUIPMENT ISSUED

Item	Laboratory serial no.
DMM	
DC power supply	

RÉSUMÉ OF THEORY

In this experiment, three different biasing circuits will be analyzed. In theory, the procedure for biasing a JFET is the same as that for a BJT. In particular, given the drain curve characteristics of the JFET and the external circuit connected to the JFET, a load line is constructed involving V_{DD}, V_{DS} and I_D. The intersection of that load line with the drain curve characteristics determines the quiescent operating point for the JFET. It is noted that the characteristics of the device is a property of the JFET; by contrast: the load line is dependent on the external circuit elements connected to the JFET. The quiescent operating point is determined by the intersection of the two curves.

In practice, JFETs, even of the same type, show considerable variation in their drain curve characteristics. As a result, manufacturers often do not publish these curves; rather, the values for the saturation current and the pinch-off voltage are given as part of the specifications. This leads to an alternative approach to determine the quiescent condition for a JFET.

To begin, the transconductance curve, which shows the relationship between V_{GS} and I_D for a particular JFET, is constructed from the saturation current, the pinch-off voltage, and Shockley's equation. Next, a bias curve is constructed sensitive to the external circuit elements connected to the JFET. The quiescent condition is determined by the intersection of the two curves.

PROCEDURE

Part 1. Fixed-Bias Network

For the fixed-bias configuration, V_{GS} will be set by an independent dc supply. The vertical lines of constant V_{GS} will intersect the transfer curve developed from Shockley's equation.

    a. Construct the network of Fig. 13.1. Insert the measured value of R_D.

    b. Set V_{GS} to zero volts and measure the voltage V_{R_R}. Calculate I_D from $I_D = V_{R_D}/R_D$ using the measured value of R_D. Since $V_{GS} = 0$ V the resulting drain current is the saturation value I_{DSS}. Record below.

(from measured) $I_{DSS} = $ _____

    c. Make V_{GS} more and more negative until $V_{R_D} = 1$ mV (and $I_D = V_{R_D}/R_D \cong 1$ µA). Since I_D is very small ($I_D \cong 0$ A), the resulting value of V_{GS} is the pinch-off voltage V_P. Record below.

Procedure

(measured) $V_P =$ _____

These values will be used throughout the experiment.

d. Using the values above for I_{DSS} and V_P, sketch the transfer curve on Fig. 13.2 using Shockley's equation.

e. If $V_{GS} = -1$ V, determine I_{D_Q} from the curve of Fig. 13.2. Show all work in Fig. 13.2. Label the straight line defined by V_{G_S} as the fixed-bias line.

(calculated) $I_{D_Q} =$ _____

f. Set $V_{GS} = -1$ V in Fig. 13.1 and measure V_{R_D}. Calculate I_{D_Q} using the measured value of R_D and insert below. This is the measured value of I_D.

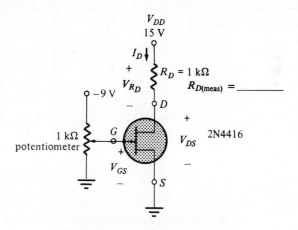

Figure 13-1 Fixed-bias circuit.

(measured) $V_{R_D} =$ _____

(from measured) $I_{D_Q} =$ _____

g. Compare the measured and calculated values of I_{D_Q}.

Part 2. Self-Bias Network

In the self-bias configuration, the magnitude of V_{GS} is defined by the product of the drain current I_D and source resistance R_S. The network bias line will start at the origin and intersect the transfer curve at the quiescent (DC) point of operation. The resulting drain current and gate-to-source voltage can then be determined from the graph by drawing a horizontal and a vertical line from the quiescent point to the axis, respectively. *Note:* The larger the

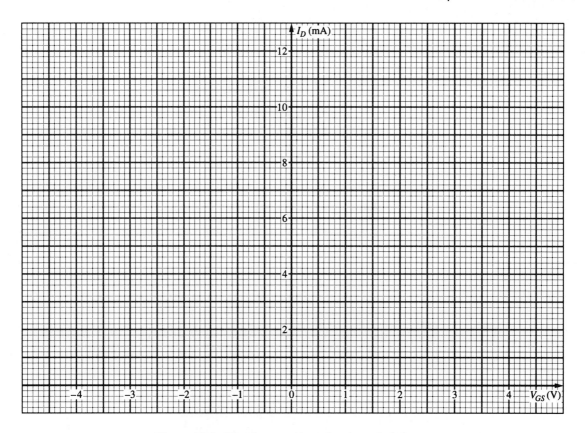

Figure 13-2 Bias lines and transfer characteristics.

source resistance, the more horizontal the bias line and the less the resulting drain current.

    a. Construct the network of Fig. 13.3. Insert the measured value of R_D and R_S.

    b. Draw the self-bias line defined by $V_{GS} = -I_D R_S$ in Fig. 13.2 and find the network Q point. Record the quiescent values of I_{D_Q} and V_{GS_Q} below. Label the straight line as the self-bias line.

(calculated) I_{D_Q} = _____

(calculated) V_{GS_Q} = _____

Procedure

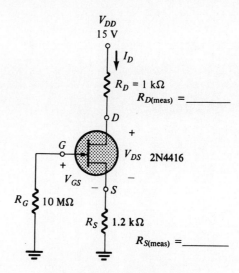

Figure 13-3 Self-bias circuit.

c. Calculate the values of V_{GS}, V_D, V_{DS}, and V_G and record below.

(calculated) V_{GS} = _____
(calculated) V_D = _____
(calculated) V_S = _____
(calculated) V_{DS} = _____
(calculated) V_G = _____

d. Measure the voltages V_G, V_{DS}, V_D, V_S, and V_G and compare with the results above using the equation

$$\% \text{ difference} = \frac{|V_{\text{meas}} - V_{\text{calc}}|}{|V_{\text{calc}}|} \times 100\% \qquad (13.1)$$

(measured) V_{GS} = _____
(measured) V_D = _____
(measured) V_S = _____
(measured) V_{DS} = _____
(measured) V_G = _____

(calculated) % (V_{GS}) = _____
(calculated) % (V_D) = _____
(calculated) % (V_S) = _____
(calculated) % (V_{DS}) = _____
(calculated) % (V_G) = _____

Part 3. Voltage-Divider-Bias Network

In the voltage-divider-bias configuration V_{GS} is determined by a voltage-divider-bias voltage and voltage drop across the source resistance. That is, for the network of Fig. 13.4,

$$V_G = \frac{R_2 V_{DD}}{R_1 + R_2}$$

and

$$V_{GS} = V_G - I_D R_S$$

a. Construct the network of Fig. 13.4. Insert the measured resistor values.

b. Using the I_{DSS} and V_P determined in Part 1, draw the voltage-divider-bias line in Fig. 13.2 and find the network Q point. Label the resulting straight line as the voltage-divider line.

   To draw the bias line determine two points as follows and then connect the two points with a straight line.

For $V_{GS} = V_G - I_D R_S$

If $I_D = 0$ mA   then $V_{GS} = V_G - (0)(R_S) = V_G$

and if $V_{GS} = 0$ V   then $I_D = \dfrac{V_G}{R_S}$

Procedure

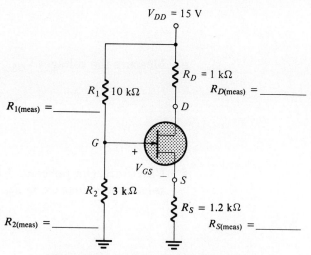

Figure 13-4 Voltage-Divider-Bias Circuit

c. Draw a straight line through the above two points and extend it until it intersects the transfer curve. The coordinates of that intersection determine the quiescent values of I_D and V_{GS}. Record them below.

(calculated) I_{D_Q} = _____

(calculated) V_{GS_Q} = _____

d. Calculate the theoretical values of V_D, V_S and V_{DS} and record below.

(calculated) V_D = _____
(calculated) V_S = _____
(calculated) V_{DS} = _____

e. Measure the voltages V_{GS_Q}, V_D, V_S, and V_{DS} and record below.

(measured) V_{GS_Q} = _____
(measured) V_D = _____
(measured) V_S = _____
(measured) V_{DS} = _____

f. Calculate the percent difference between the measured and calculated values using Eq. (13.1).

(calculated) % (V_{GS_Q}) = _____
(calculated) % (V_D) = _____
(calculated) % (V_S) = _____
(calculated) % (V_{DS}) = _____

g. Calculate I_{D_Q} from the measured voltages of step **e** and compare to the value determined in step **c**. I_{D_Q} can be found using

$$I_{D_Q} = \frac{V_{DD} - V_D}{R_D}$$

and the measured values of V_D and R_D. Record below and calculate the % difference.

Problems and Exercises

(measured) I_{D_Q} = _____

(calculated) % (I_{D_Q}) = _____

Part 4. Computer Exercise

a. Perform a DC analysis of the network of Fig. 13.4 using PSpice, MiroCap II or other appropriate software package. That is, find V_{GS_Q}, I_{D_Q}, V_D, V_S, and V_{DS}. The input file is the following assuming I_{DSS} = 8 mA and V_P = −4 V. For your input file insert the measured values of I_{DSS} and V_P. Remember that VTO = V_P and BETA = $I_{DSS}/|V_P|^2$.

```
VOLTAGE-DIVIDER CONFIGURATION OF FIG. 13.4
VDD    2   0        DC    15V
R2     1   0        3K
R1     2   1        10K
RD     2   3        1K
RS     4   0        1.2K
J1     3   1    4        JFET
.MODEL JFET NJF(VTO = -4V BETA = 5E - 4)
.DC VDD    15V   15V   1V
.OPTIONS NOPAGE
.PRINT DC V(1,4), I(RD), V(3), V(4), V(3,4)
.END
```

b. Repeat the above analysis for the self-bias circuit of Fig. 13.3.

c. How do the results of steps 4a and 4b compare with the measured values of the experiment?

Problems and Exercises

1. a. Record your values of I_{DSS} and V_P and the values for both quantities from two other squads.

(your squad) I_{DSS} = _____, V_P = _____
(other squads) I_{DSS} = _____, V_P = _____
I_{DSS} = _____, V_P = _____

b. Is the range more than you expected for a specified JFET? What effect will the range of values have on the design process?

2. For the self-bias configuration what is the effect of increasing values of R_S on the resulting Q point? That is, does an increasing value of R_S result in an increase or decrease in the level of I_{D_Q}? What is the effect of an increasing value of R_S on V_{GS_Q}? Explain in some detail.

3. What value of source resistance (R_S) would make the quiescent drain current equal to 1/2 the saturation level (I_{DSS}) for the self-bias configuration? Use the parameter values of Fig. 13.3 and your level of I_{DSS} and V_P.

(calculated) $R_S =$ _____

4. What value of source resistance (R_S) would make the quiescent drain current I_{D_Q} equal to 1/2 the saturation level (I_{DSS}) for the voltage-divider configuration of Fig. 13.4? Use the parameter values of Fig. 13.4 and your level of I_{DSS} and V_P.

(calculated) $R_S =$ _____

Name _____
Date _____
Instructor _____

EXPERIMENT 14
Design of JFET Bias Circuits

OBJECTIVES

1. To design a self-bias JFET circuit for given bias conditions.
2. To design a voltage-divider-bias JFET circuit for given bias conditions.
3. To test both of these circuits, and if necessary, redesign them.

EQUIPMENT REQUIRED

Instruments

Dual-trace oscilloscope
DMM
DC power supply
9 V battery with snap-on leads

Components

Resistors

(1) 1 kΩ
(1) 1-kΩ potentiometer

Since this is a design experiment a number of resistor values will have to be determined and requested from the stockroom.

Transistors

(1) 2N4416 JFET or equivalent

163

Exp. 14 / Design of JFET Bias Circuits

EQUIPMENT ISSUED

Item	Laboratory serial no.
Oscilloscope	
DMM	
DC power supply	
Signal generator	

RÉSUMÉ OF THEORY

Like the BJT, the junction field-effect transistor can operate in three modes: cutoff, saturation, and linear. The physical characteristics of the JFET and the external circuit elements connected to it determine the mode of operation. In this experiment, the JFET is biased in the linear mode in accordance with a given set of circuit specifications.

JFETs, even of the same type, show a considerable variation in their characteristics. In consequence, manufacturers rarely publish the drain characteristics of JFETs but simply specify both the saturation current I_{DSS} and the pinch-off voltage V_P. The designer can construct the transfer characteristic from these two values and any intermittent values of I_D and V_{GS} from Shockley's equation.

The transfer curves together with the operating specifications will be used to determine needed values for the various circuit elements to be used in the two bias designs. In both cases, a procedure will be suggested that will place the Q-point at or near the specified DC operating conditions and will allow for a specified AC signal to be amplified without distortion. Upon completion of the design, the experimenter will construct the actual circuit and take the DC measurements to ensure correct operation of the circuit. In case the design does not meet the specifications, a corrective procedure is suggested.

PROCEDURE

Part 1. Determining I_{DSS} and V_P

This part of the experiment will determine I_{DSS} and V_P for the JFET to be employed in the design process of this experiment.

    a. Construct the network of Fig. 14.1. Insert the measured value of R_D.

    b. Set V_{GS} to zero volts and measure V_{R_D}. Calculate $I_D = I_{DSS} = V_{R_D}/R_D$ using the measured resistance value and record below.

(measured) I_{DSS} = _____

    c. Make V_{GS} more and more negative until V_{R_D} = 1 mV (and $I_D = V_{R_D}/R_D \cong 1\ \mu A$). Since I_D is small ($I_D \cong 0$ A), the resulting value of V_{GS} is the pinch-off voltage V_P. Record below.

(measured) V_P = _____

Procedure

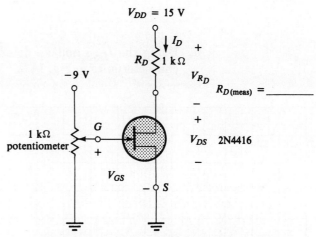

Figure 14-1 Network to determine I_{DSS} and V_P.

Part 2. Self-bias Circuit Design

This part of the experiment will determine R_D and R_S for a self-bias configuration of Fig. 14.2 to establish a Q point at $I_{D_Q} = I_{DSS}/2$, $V_{DS_Q} = V_{DS_{max}}/2$ with $V_{DD} = 2 V_{DS_Q}$.

    a. Using the specified Q-point and the results of Part 1 with the constraint that $V_{DS_{max}} = 30$ V calculate I_{D_Q}, V_{DS_Q} and V_{DD}.

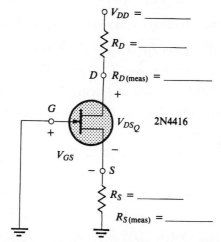

Figure 14-2 Self-bias configuration to be designed.

(calculated) I_{D_Q} = _____

(calculated) V_{DS_Q} = _____

(calculated) V_{DD} = _____

b. Using the I_{DSS} and V_P determined in Part 1 sketch the transfer characteristics in Fig. 14.3.

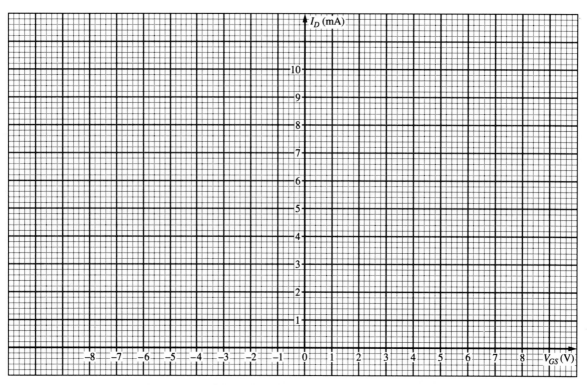

Figure 14-3 Transfer characteristic for the 2N4416 JFET.

c. The choice of $I_{D_Q} = I_{DSS}/2$ will permit a maximum swing in collector current for the AC domain. Draw a horizontal line from $I_{DSS}/2$ on the I_C axis to the transfer curve of Fig. 14.3 and label the intersection as the Q point to define the self-bias line. Label the resulting line as the self-bias line for future reference.

d. Determine the value of R_S from

$$R_S = \left| \frac{\Delta V}{\Delta I} \right| = \left| \frac{\Delta V_{GS}}{\Delta I_D} \right|$$

where | | specifies the magnitude of the quantity and Δ the change in each quantity from the origin to the Q point.

Procedure

(calculated) $R_S =$ _____

Determine the closest commercial value (available in the laboratory) to the calculated R_S and insert in the space provided on Fig. 14.2. In addition, insert the calculated value for V_{DD} from Part 2(a).

(commercial value) $R_S =$ _____

Insert the commercial and measured value of R_S in the space provided on Fig. 14.2. In addition, insert the calculated value for V_{DD} from Part 2(a).

e. The level of R_D will now be determined by an application of Kirchhoff's voltage law to the output circuit of Fig. 14.2 followed by the use of Ohm's law. For the output circuit of Fig. 14.2,

$$V_{R_D} = V_{DD} - V_{DS_Q} - V_{R_S}$$

Determine $V_{R_S} = I_{D_Q} R_S$ (where R_S is the measured value) and substitute the levels of V_{DS_Q} and V_{DD} from Part 2(a) to determine V_{R_D}.

(calculated) $V_{R_D} =$ _____

Determine R_D using I_{D_Q} from Part 2(a) and Ohm's law.

(calculated) $R_D =$ _____

Determine the closest commercial value (available in the laboratory) to R_D and record below.

(commercial value) $R_D =$ _____

Insert the commercial value and the measured value of R_D in the space provided in Fig. 14.2.

f. Using the commercial values of R_D and R_S and the calculated value of V_{DD}, construct the network of Fig. 14.2. Energize the network and measure V_{DS_Q} and V_{R_D}. Use the measured value of R_D to calculate I_{D_Q}. Record the levels of V_{DS_Q} and I_{D_Q} below.

(measured) $V_{DS_Q} =$ _____

(measured) $I_{D_Q} =$ _____

Record the design levels of I_{D_Q} and V_{DS_Q} calculated in Part 2(a).

(calculated) V_{DS_Q} = _____

(calculated) I_{D_Q} = _____

g. Are you pleased with the results of Part 2(f)? If not, how would you improve the design? Be specific.

h. Borrow the 2N4416 JFET from the group next to you and insert in your network using the values of R_D, R_S, and V_{DD} determined above. Measure the resulting levels of V_{DS_Q} and V_{R_D} and calculate I_{D_Q} as above. Insert the results below.

(measured) V_{DS_Q} = _____

(measured) I_{D_Q} = _____

In addition, record the levels of I_{DSS} and V_P for the borrowed JFET transistor.

(borrowed JFET) I_{DSS} = _____

(borrowed JFET) V_P = _____

How do the values of V_{DS_Q} and I_{D_Q} compare with the specified design values of Part 2(a) for this borrowed transistor? Use the recorded values of I_{DSS} and V_P for the other JFET to help explain any variations.

What do the results obtained above tell you about the design process when using JFETs of the same nameplate data?

The specification sheet for the 2N4416 JFET specifies a range of 5 mA to 15 mA for I_{DSS} and −1 V to −6 V for V_P. Are the levels of I_{DSS} and V_P obtained for your JFET and the borrowed JFET in this range? Are the average values of 10 mA and −3.5 V good choices for a first design effort if I_{DSS} and V_P are unknown quantities?

Part 3. Voltage-divider circuit Design

This part of the experiment will determine the value of R_D, R_S, R_1, and R_2 for the voltage-divider configuration of Fig. 14.4. The Q-point is to be established at

$$I_{D_Q} = 4 \text{ mA}$$

and $V_{DS_Q} = 8 \text{ V}$

Additional specifications:

$$V_{DD} = 20 \text{ V}$$
$$R_2 = 10 R_S$$

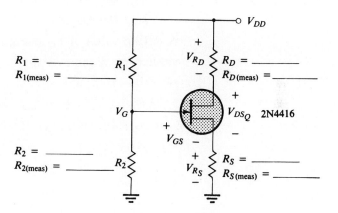

Figure 14-4 Voltage-divider configuration to be designed.

a. First locate the Q-point on the transfer curve of Fig. 14.3 by drawing a horizontal line from $I_{D_Q} = 4$ mA on the vertical axis to the transfer curve. At the Q-point draw a vertical straight line to determine the corresponding value of V_{GS}.

(calculated) $V_{GS} = $ _____

The network equation for V_{GS} has the following format:

$$V_{GS} = V_G - I_D R_S \tag{14.1}$$

A design decision must now be made. Both V_G and R_S are undefined quantities. However we know V_G must be a positive voltage but too small a level of voltage will result in a very low resistance level for R_S. A low level of R_S will in turn result in an undesirably high level of DC dissipation during operation. The high range of V_G is limited by the available supply. As a compromise, let us choose V_G to be equal to the magnitude of the pinch-off voltage V_P. That is

$$V_G = |V_P| \qquad (14.2)$$

The level of R_S can then be determined using equation 14.1.

Using the defined quiescent conditions, the resulting level of V_{GS}, and the V_P for your 2N4416 transistor, calculate the required level of R_S.

(calculated) $R_S =$ _____

Determine the closest commercial value (available in the laboratory) to R_S and insert below.

(commercial value) $R_S =$ _____

Insert the commercial and measured value for R_S in Fig. 14.4.

Using the measured value of R_S, V_{DS_Q} and I_{D_Q} determine the resulting value of V_G using Eq. 14.1. Is it relatively close to the value determined by V_P?

(calculated) $V_G =$ _____

c. For the output circuit

$$V_{R_D} = V_{DD} - V_{DS_Q} - V_{R_S}$$

Calculate V_{R_S} from $V_{R_S} = I_{D_Q} R_S$ (using the measured resistance value) and substitute the specified values of V_{DD} and V_{DS_Q} to determine V_{R_D}.

Procedure   171

(calculated) V_{R_D} = _____

Determine R_D using I_{D_Q} and Ohm's law.

(calculated) R_D = _____

Determine the closest commercial value (available in the laboratory) to R_D and insert below.

(commercial value) R_D = _____

Insert the commercial and measured value of R_D in Fig. 14.4.

d. R_1 and R_2 will now be determined using the following equation:

$$V_G = \frac{R_2 V_{DD}}{R_1 + R_2} \qquad (14.3)$$

and the specification

$$R_2 = 10 R_S \qquad (14.4)$$

Using the commercial value of R_S, calculate the level of R_2 using Eq. 14.4.

(calculated) R_2 = _____

Determine the closest commercial value (available in the laboratory) to R_2.

(calculated) R_2 = _____

Insert the commercial and measured values for R_2 in Fig. 14.4.

Using the specified value of V_{DD}, the calculated value of V_G (from the measured value of R_S) and the commercial value of R_2, calculate the value of R_1 using Eq. 14.3.

(calculated) R_1 = _____

Determine the closest commercial value (available in the laboratory) to R_1 and insert below.

(commercial value) R_1 = _____

Insert the commercial and measured values of R_1 in Fig. 14.4.

e. Using the commercial values of R_D, R_S, R_1, and R_2 and the specified value of V_{DD}, construct the network of Fig. 14.4. Energize the network and measure V_{DS_Q} and V_{R_D}. Using the measured value of R_D, calculate I_{D_Q}. Record the levels of V_{DS_Q} and I_{D_Q} below.

(measured) V_{DS_Q} = _____

(measured) I_{D_Q} = _____

Record the specified design values for V_{DS_Q} and I_{D_Q} below.

(specified) V_{DS_Q} = _____

(specified) I_{D_Q} = _____

f. Using the following equations determine the percent difference between the specified and measured values of I_{D_Q} and V_{DS_Q}.

$$\% I_{D_Q} = \frac{|I_{D_{Q(\text{specified})}} - I_{D_{Q(\text{measured})}}|}{|I_{D_{Q(\text{specified})}}|} \quad (14.5)$$

$$\% V_{DS_Q} = \frac{|V_{DS_{Q(\text{specified})}} - V_{DS_{Q(\text{measured})}}|}{|V_{DS_{Q(\text{specified})}}|} \quad (14.6)$$

(calculated) $\% I_{D_Q}$ = _____

(calculated) $\% V_{DS_Q}$ = _____

Are you satisfied with your design effort? Be specific.

Procedure

g. If the percent difference is more than 10%, how would you improve the design? Take careful note of your voltage-divider bias line on Fig. 14.3 when you consider alternative designs.

Make adjustments in your design to reduce the percent differences in I_{D_Q} and V_{DS_Q} to less than 10%. Record the new values of R_D, R_S, R_1, and R_2 below. In addition, record the new values of I_{D_Q} and V_{DS_Q} and calculate the percent differences with the specified levels.

(commercial value) $R_D =$ _____

(commercial value) $R_S =$ _____

(commercial value) $R_1 =$ _____

(commercial value) $R_2 =$ _____

(measured) $I_{D_Q} =$ _____

(measured) $V_{DS_Q} =$ _____

(calculated) % $I_{D_Q} =$ _____

(calculated) % $V_{DS_Q} =$ _____

h. Again borrow the same 2N4416 JFET from the group next to you and insert in your network using the resistors chosen for your design. Measure the resulting levels of V_{DS_Q} and V_{R_D} and calculate I_{D_Q} using the measured resistor value. Insert the results below.

(measured) V_{DS_Q} = _____

(measured) I_{D_Q} = _____

In addition, record the levels of I_{DSS} and V_P for the borrowed JFET.

(borrowed JFET) I_{DSS} = _____

(borrowed JFET) V_P = _____

How do the values of V_{DS_Q} and I_{D_Q} compare with the specified design levels using the borrowed JFET? Use the recorded values of I_{DSS} and V_P for the borrowed JFET to help explain any variations.

How do the overall results for this configuration compare to that obtained for the fixed-bias configuration when the JFETs are interchanged? Does one configuration appear to be less sensitive to changes in JFETs of the same nameplate nomenclature?

Problems and Exercises

1. Using the average values of I_{DSS} = 10mA and V_P = −3.5 V from the specification sheets for the 2N4416 JFET, redesign the self-bias configuration of Fig. 14.2 for the specified Q-point.

(commercial value) R_D = _____

(commercial value) R_S = _____

Problems and Exercises

Are the values of R_D and R_S within 20% of the design values of Part 2? Comment accordingly.

2. Using the average values of $I_{DSS} = 10$ mA and $V_P = -3.5$ V from the specification sheets for the 2N4416 JFET, redesign the voltage-divider configuration of Fig. 14.4 for the specified Q-point.

(commercial value) $R_D =$ _____
(commercial value) $R_S =$ _____
(commercial value) $R_1 =$ _____
(commercial value) $R_2 =$ _____

Are the values or R_D, R_S, R_1, and R_2 within 20% of the design values of Part 3? Comment accordingly.

3. How else may a designer cope with the variation in I_{DSS} and V_P for JFETs with the same nameplate nomenclature?

Name _____
Date _____
Instructor _____

EXPERIMENT 15

Compound Configurations

OBJECTIVES

1. To measure the bias voltages and currents of multistage systems.
2. To demonstrate the independence of the DC voltages and currents of one stage in a capacitively coupled system with the DC voltages and currents of any other stage of the system.
3. To measure the bias voltages and currents of a multistage dc coupled system.

EQUIPMENT REQUIRED

Instruments

DMM

Supplies

DC power supply
9 V battery with snap-on leads

Components

Resistors

(1) 470 Ω
(2) 1 kΩ
(1) 1.2 kΩ
(1) 2.4 kΩ
(1) 2.7 kΩ
(1) 4.7 kΩ

(2) 15 kΩ
(1) 1 kΩ potentiometer

Capacitor

(1) 0.1-µF

Transistors

(2) 2N3904 BJT
(1) 2N4416 JFET

EQUIPMENT ISSUED

Item	Laboratory serial no.
DMM	
DC power supply	

RÉSUMÉ OF THEORY

Typical electronic amplifying systems consist of several transistor stages connected together. The amplification purpose dictates the nature of the interconnection between the stages. If an amplifier is required to amplify a signal containing frequencies well above 0 Hz, the method of coupling most commonly employed is ac coupling. It consists of connecting a capacitor between the collector of one stage and the base of the next stage. In this fashion, the ac component of the collector output voltage is connected into the base of the next stage, while the dc component of the collector voltage is blocked from reaching that base due to the capacitor. In effect, relative to any dc voltages and currents, stages so coupled are isolated from each other. This makes the dc analysis of even the most complex systems relatively easy since each stage can be analyzed independently. In this experiment, the dc biasing levels of various stages of an amplifier are measured. It is then demonstrated that the capacitively coupled stages do not affect each other's dc voltages and currents. Both an exchange of position of the stages in the system, and a change in the biasing network of a transistor are used in that demonstration.

The second coupling system investigated during this experiment is a dc-coupled system. Such systems are used when very low frequency components of a signal and even its dc component need to be amplified. A direct connection is made between the collector of a stage and the base of the next. Here it will be demonstrated that any change in the dc voltages and currents in one stage affects the dc voltages and currents in another stage. A technique of changing the bias network on one stage will be used to show the DC dependence of the two stages used in this experiment.

The third compound bias circuit will include a BJT-JFET combination to demonstrate analysis techniques employed for such configurations. The coupling will be direct-coupled to permit a full investigation of the interaction between active devices.

Procedure

PROCEDURE

Part 1. Determining the BJT(β) and JFET(I_{DSS} and V_P) parameters

This part of the experiment will determine the BJT and JFET parameters to be employed in the analysis of each compound configuration.

a. To determine the β for each BJT transistor construct the network of Fig. 15.1 and insert the measured resistor values.

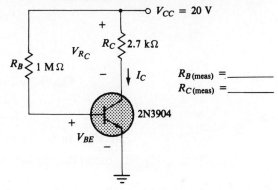

Figure 15-1 Determining β.

Energize the network and measure the voltages V_{BE} and V_{R_C}. Using the measured values of R_B and R_C, calculate the levels of I_B and I_C, using the equations $I_B = (V_{CC} - V_{BE})/R_B$ and $I_C = V_{R_C}/R_C$. Then calculate β from $\beta = I_C/I_B$ and insert below.

(measured) $\beta_1 = $ _____

Replace the 2N3904 with the other BJT transistor and determine its level of β. Insert its level below. Throughout the experiment be sure you can identify which transistor has which level of β.

(measured) $\beta_2 = $ _____

b. To determine I_{DSS} and V_P for the JFET construct the network of Fig. 15.2 and insert the measured value of R_D.

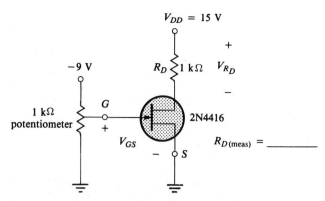

Figure 15-2 Determining I_{DSS} and V_P.

Set V_{GS} to 0 V and measure V_{R_D}. Calculate $I_D = I_{DSS} = V_{R_D}/R_D$ using the measured resistor value and record below.

(measured) I_{DSS} = _____

Make V_{GS} more and more negative until $V_{R_D} = 1$ mV (and $I_D = V_{R_D}/R_D \cong 1$ µA). Since I_D is small ($I_D \cong 0$ A), the resulting value of V_{GS} is the pinch-off voltage V_P. Record below.

(measured) V_P = _____

Part 2. Capacitive-Coupled Multistage System with Voltage-Divider Bias

In this part, bias voltages and currents of a capacitively-coupled two-stage amplifier system are measured. The dc isolation between the two stages will be demonstrated.

    a. Construct the circuit of Fig. 15.3 using a 2N3904 transistor for each stage.

Procedure

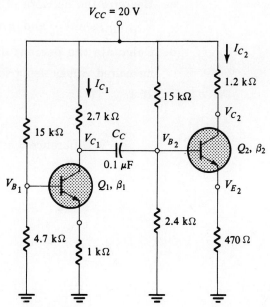

Figure 15-3 AC-coupled multistage amplifier.

b. Using the β values determined in Part 1, calculate the voltage levels V_{B_1}, V_{C_1}, V_{B_2}, and V_{C_2} using commercial values. Keep in mind that the coupling capacitor C_C will assume the "open-circuit" state for dc conditions. Insert the results in Table 15.1.

TABLE 15.1

	V_{B_1}	V_{C_1}	V_{B_2}	V_{C_2}
Calculated Values				
Measured Values				
% Difference				

c. Energize the network of Fig. 15.3 and measure the voltages V_{B_1}, V_{C_1}, V_{B_2}, and V_{C_2} and insert in Table 15.1.

d. Calculate the percent differences between the calculated and measured values using the following equation and insert in Table 15.1.

$$\% \text{ Difference} = \frac{|V_{(calc)} - V_{(meas)}|}{|V_{(calc)}|} \times 100\% \tag{15.1}$$

e. Even though commercial resistor values were employed, are the percent differences in general less than 10%? If not, can you comment on why the difference was so large?

f. Compare the measured values of V_{C_1} and V_{B_2}. Do they confirm the fact that the capacitor C_C assumes an open-circuit state for dc conditions? In other words, for dc conditions, are the two voltage-divider configurations isolated?

Part 3. DC-Coupled Multistage Systems

In this part the bias voltages of a dc-coupled two-stage transistor amplifier will be calculated, measured and compared. The primary purpose is to demonstrate that the dc levels of one stage will have a direct effect on the dc levels of the other stage.

Procedure

a. Construct the network of Fig. 15.4 using the 2N3904 transistors.

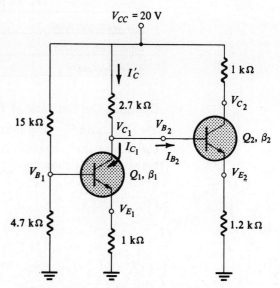

Figure 15-4 DC-coupled multistage amplifier.

b Using the β values determined in Part 1, calculate the voltage levels V_{B_1}, V_{C_1}, V_{B_2}, and V_{C_2} using commercial values. In this case, proceed by first finding V_{B_1}, then V_{E_1}, I_{E_1} and I_{C_1} followed by V_{C_1} assuming $I'_C \cong I_{C_1} \gg I_{B_2}$. Once $V_{C_1} = V_{B_2}$ is known V_{E_2} and the remaining unknowns can be found. Insert the results in Table 15.2.

TABLE 15.2

	V_{B_1}	V_{C_1}	V_{B_2}	V_{C_2}
Calculated Values				
Measured Values				
% Difference				

c. Energize the network of Fig. 15.4 and measure the voltages V_{B_1}, V_{C_1}, V_{B_2}, and V_{C_2} and insert in Table 15.2.

d. Calculate the percent differences between the calculated and measured values using Eq. 15.1 and insert in Table 15.2.

e. Even though commercial resistor values were employed, are the percent differences in general less than 10%? If not, can you comment on why the difference was so large?

f. Compare the measured values of V_{C_1} and V_{B_2}. Are they equal as expected? Comment on how the dc voltage levels of one stage directly affected the dc voltage levels of the other stage.

Part 4. A BJT-JFET Compound Configuration

A compound configuration with both BJT and JFET transistors will now be examined from a DC viewpoint. The configuration of Fig. 15.5 is direct-coupled resulting in a direct link in DC levels between the two transistors.

a. Construct the network of Fig. 15.5 using a 2N3904 BJT transistor and the 2N4416 JFET transistor.

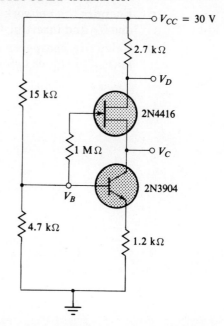

Figure 15-5 BJT-JFET Compound Configuration.

b. Using the β, I_{DSS} and V_P levels determined in Part 1, calculate the dc levels of V_B, V_C and V_D using commercial values. In this case initiate your analysis by first finding V_B and then the level of I_C. Then determine the level of V_D. Using Shockley's equation find the level of V_{GS} followed by V_C. Insert the results in Table 15.3.

TABLE 15.3

	V_B	V_D	V_C
Calculated Values			
Measured Values			
% Difference			

c. Energize the network of Fig. 15.5 and measure the voltages V_B, V_D and V_C and insert in Table 15.3.

d. Calculate the percent differences between the calculated and measured values using Eq. 15.1 and insert in Table 15.3.

e. Even though commercial resistor values were employed are the percent differences in general less than 10%? If not, can you comment on why the difference was so large?

f. Determine the voltage V_{GS} from the measurements of Table 15.3. How does it compare to your calculated value from Part 4(**b**)?

(measured) V_{GS} = _____
(calculated) V_{GS} = _____

Procedure

g. Determine the voltage V_{R_D} from measured values and calculate the drain current from Ohm's law using the commercial resistor value. How does the measured value of I_D compare to the calculated value of Part 4(b)?

(measured) $I_D = $ _____
(calculated) $I_D = $ _____

h. Using $V_{BE} = 0.7$ V calculate the voltage V_E from the measured values of Table 15.3 and then calculate I_C using the commercial resistor value. How does the measured value of I_C compare to the measured value of I_D in Part 4(g)?

(measured) $I_C = $ _____

Problems and Exercises

1. **a.** For the network of Fig. 15.3 how will the level of V_B and V_C for each transistor change if the two voltage-divider configurations are interchanged?

   b. Will the level of V_B and V_C for each transistor of Fig. 15.3 change if the resistive components maintain their current positions and the transistors are interchanged? Why?

2. Will there be a significant change in the level of V_{E_2} for the network of Fig. 15.4 if the resistive components maintain their current positions and the transistors are interchanged? Support your conclusions with numerical calculations.

3. Remove the 1 MΩ resistor and interchange the positions of the BJT and JFET of Fig. 15.5. Calculate the resulting level of V_B, V_D, and V_C and compare to the levels of Part 4(b). Have they changed considerably? Was the change in level expected? Why?

Name _____
Date _____
Instructor _____

EXPERIMENT 16

Measurement Techniques

OBJECTIVES

1. To measure the ac and dc amplitudes of a waveform with an oscilloscope.
2. To measure the ac and dc amplitudes of a waveform with a digital multimeter.
3. To measure the period and frequency of periodic waves with an oscilloscope.
4. To measure the frequency of periodic waves with a frequency counter.
5. To measure the phase shift between two sinusoidal waves with an oscilloscope.
6. To study the effect of instrument loading on the voltage measurements in a circuit.

EQUIPMENT REQUIRED

Instruments

Dual-trace oscilloscope
 (Single-trace will limit investigation to Parts 1, 2, and 4)
DMM
Frequency counter

Supplies

DC power supply
Signal generator

Components

Resistors

(2) 1-kΩ

(1) 2-kΩ
(1) 3.9-kΩ
(2) 1-MΩ

Capacitor

(1) 0.1 µF

EQUIPMENT ISSUED

Item	Laboratory serial no.
Oscilloscope	
DMM	
DC power supply	
Signal generator	
Frequency counter	

RÉSUMÉ OF THEORY

This experiment will be an introduction to the measuring instrumentation commonly used to measure dc and ac quantities. Specifically, the oscilloscope and the digital multimeter will be used to measure both the ac and dc components of a voltage waveform. The oscilloscope is basically a voltage-measuring device. It measures the amplitude of any periodic voltage in terms of its peak-to-peak values. By contrast, the digital multimeter measures the rms value of a periodic wave. Note, however, that some digital multimeters measure the rms value of a sinusoidal wave only.

The oscilloscope can also be used to measure the period, and consequently the frequency, of a periodic wave. In case of a sine wave, it will measure the frequency of that wave; if a pulse is applied, it will allow for the determination of its fundamental frequency. The frequencies determined from the oscilloscope measurements will be compared with those made with a frequency counter.

It is important in the use of a particular measuring instrument to note its possible effect on the measurements taken. To demonstrate this, a circuit is used which has a low impedance compared to the input impedance of the oscilloscope. Thus the circuit voltage measured approaches its theoretical value. However, when the circuit impedance is changed so that it more nearly approaches that of the oscilloscope, serious measurement errors are introduced. To overcome these errors, 10:1 test probes are used.

PROCEDURE

Part 1. AC and DC Voltage Amplitude Measurements

a. Construct the circuit of Fig. 16.1. Insert the measured resistor values.

Procedure

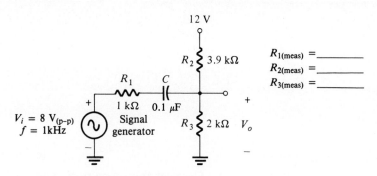

Figure 16-1 AC and DC voltage measurements.

b. Connect the oscilloscope to measure the voltage V_i. For the channel being used set the AC-GND-DC switch to the GND position and set the horizontal line in the middle of the screen. Then return the AC-GND-DC switch to the AC position.

c. Set the vertical sensitivity to 1 V/cm and adjust the amplitude control of the signal generator until $V_i = 8$ V_{p-p} at a frequency of 1 kHz. Use a horizontal sensitivity of 0.2 ms/cm.

d. Set the DC supply to 12 V using the DMM.

The network is now established with both an AC and DC supply.

DC MEASUREMENTS

Both the oscilloscope and DMM will now be used to measure the DC levels of Fig. 16.1.

e. Calculate the expected DC voltage level at V_o using the measured resistor values.

(calculated) $V_o =$ _____

f. Use the DMM to measure the DC level of V_o.

(measured) $V_o =$ _____

Determine the percent difference between the calculated and measured values using the following equation:

$$\% \text{ Difference} = \frac{|V_{o(\text{calc})} - V_{o(\text{meas})}|}{|V_{o(\text{calc})}|} \times 100\% \qquad (16.1)$$

(calculated) % Difference = _____

g. Connect the scope to V_o and set the AC-GND-DC switch to the DC position. Using a sensitivity of 1 V/cm determine the shift (in volts) in the positive peak value (referenced to 0 V) from the established in Part1(**c**).

(measured) shift in V_o = _____

Was the shift up or down from the center of the screen? What does the shift tell us about the polarity of V_o.

How does the measured shift with the oscilloscope compare with that measured with the DMM?

Is the scope or DMM more accurate for this type of reading? Why?

AC Measurements

Both the oscilloscope and DMM will now be used to measure the ac levels of Fig. 16.1.

h. Calculate the rms value of the applied voltage V_i.

(calculated) $V_{i(\text{rms})}$ = _____

i. Calculate the expected rms voltage V_o for the network of Fig. 16.1 at a frequency of 1 kHz, using measure resistor values. Be aware

Procedure

the reactance of the capacitor must be determined and the vector relationship between resistive and reactive elements employed in the determination. For the ac analysis the 12 V supply can be set to zero volts (superposition applies to the DC/AC analysis of the network) resulting in a parallel arrangement for R_2 and R_3.

(calculated) $V_{o(rms)}$ = _____

j. Use the DMM to measure the rms value of V_o.

(measured) $V_{o\ (rms)}$ = _____

Determine the percent difference between the calculated and measured values using Eq. 16.1.

(calculated) % Diff. = _____

k. Connect the oscilloscope to measure V_o and set the AC-GND-DC switch the AC position. Using an appropriate vertical and horizontal sensitivity determine the peak-to-peak value of V_o.

(measured) $V_{o(p\text{-}p)}$ = _____

Calculate the rms value of V_o

(measured) $V_{o(rms)}$ = _____

Determine the percent difference between the calculated and measured values using Eq. 16.1.

(calculated) % Difference = _____

l. Are you satisfied that both the oscilloscope and DMM can effectively measure the rms values of sinusoidal waveforms? Why?

Part 2. Measurements of the Periods and Fundamental Frequencies of Periodic Waveforms

In this part of the experiment the oscilloscope will be used to measure the period and frequency of a sinusoidal waveform.

    a. Hook up the signal generator directly to a vertical channel of the oscilloscope. Set the frequency dial between 1 and 2 kHz *without* taking the time to carefully read the scale and determine which frequency was chosen. Adjust the amplitude control until an 8 $V_{p\text{-}p}$ signal is obtained on the screen.

    An 8 $V_{p\text{-}p}$ sinusoidal signal of unknown frequency is now displayed on the screen. The following is the general procedure to determine the period and frequency of a waveform.

    b. Adjust the horizontal sensitivity until one or two complete cycles of the waveform are displayed on the screen. Record the chosen horizontal sensitivity below.

Horizontal sensitivity = _____

    c. Measure the number of divisions (including fractional parts) encompassed by one full cycle of the waveform on the screen.

Number of divisions = _____

    d. Calculate the period of the waveform by multiplying the horizontal sensitivity by the number of divisions.

Period (T) = _____

    e. The frequency can then be determined using the relationship $f = 1/T$. Calculate the frequency.

Frequency (f) = _____

Procedure

f. Now that the frequency is known compare it to the frequency set on the signal generator. Record the set frequency below.

f (dial setting) = _____

g. If the calculated frequency and frequency of the signal generator do not match can you offer a reason for the difference?

h. Hook up the frequency counter to the output voltage terminals and record the displayed frequency.

f (counter) = _____

i. Is the frequency displayed by the counter closer to the frequency calculated using the scope or determined from the dial of the signal generator? Assuming the counter is our best measurement does a scope or dial setting usually display a more accurate reading of the frequency?

Part 3. Phase-Shift Measurements

a. Construct the network of Fig. 16.2. Insert the measured resistor value.

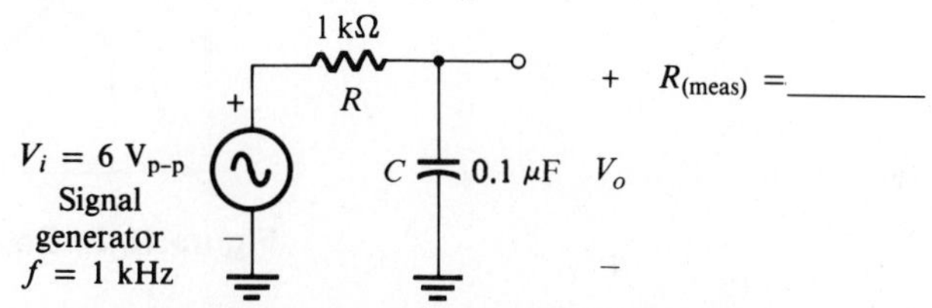

$R_{(meas)}$ = _____

$V_i = 6$ V$_{p-p}$
Signal generator
$f = 1$ kHz

$R = 1$ kΩ
$C = 0.1$ μF
V_o

Figure 16-2 Phase-shift measurements.

b. Determine the rms value of the 6 V_{p-p} signal applied to the input.

(calculated) $V_{i\,(rms)} =$ _____

c. Assuming $V_i = V_i \angle 0°$ determine $V_o \angle \theta$ at a frequency of 1 kHz.

(calculated) $V_{o(rms)} =$ _____
(calculated) $V_{o(p-p)} =$ _____
$\theta =$ _____

The angle θ is the phase angle between V_i and V_o.

d. Hook up V_i to channel 1 of the oscilloscope and establish V_i as a 6 V_{p-p}, 1 kHz sinusoidal signal balanced (using the AC-GND-DC switch) above and below the center line on the screen using a vertical sensitivity of 1 V/cm. Adjust the waveform so the intersection of the positive slope with the center line occurs at the intersection of one of the vertical grid lines as shown in Fig. 16.3.

e. Hook up channel 2 to V_o and, using the same vertical sensitivity of 1 V/cm, superimpose V_o on V_i. Be sure both V_i and V_o are balanced above and below the center line using the GND position of the AC-GND-DC switch for each channel.

f. Count the number of horizontal divisions between positive slopes of V_o and V_i as shown in Fig. 16.3 and label the result A. The separation A represents the phase shift between V_o and V_i.

A (number of divisions) = _____

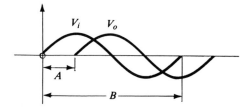

Figure 16-3 Determining the phase shift.

g. Count the number of divisions encompassed by one full cycle of the waveforms and label the result B (note Fig. 16.3).

B (number of divisions) = _____

h. The phase angle in degrees can then be determined using the following equation:

$$\theta = \frac{A}{B} \times 360° \qquad (16.2)$$

Using Eq. 16.2 calculate the phase angle between V_o and V_i for the network of Fig. 15.2.

(measured) $\theta =$ _____

How does the phase angle measured in Part 3 (**h**) compare to the phase angle calculated in Part 3 (**c**)?

i. How does the peak-to-peak value of V_o compare to the calculated value of Part 3 (**c**)?

j. If V_o crosses the axis with a positive slope to the right of V_i, V_o lags V_i by the angle θ. For the network of Fig. 16.2 does V_o lead or lag V_i? Is the result expected? Why?

k. The phase relationship between V_i and V_R can be obtained by interchanging the positions of the capacitor and resistor. The change in location is required to insure a common ground between waveforms viewed on the scope.

Interchange the positions of the resistor and capacitor of Fig. 16.2 and calculate the magnitude and angle of V_R assuming $V_i = V_i \angle 0°$.

(calculated) $V_{R(rms)}$ = _____
(calculated) $V_{R(p-p)}$ = _____
θ = _____

1. Use the oscilloscope to measure the magnitude of V_R and V_i. Also indicate if V_o leads or lags V_i.

(measured) $V_{R\ (p-p)}$ = _____
(measured) $V_{i\ (p-p)}$ = _____
θ = _____
lead or lag? _____

How do the measured and calculated results compare?

Part 4. Loading Effects

a. Construct the network of Fig. 16.4. Insert the measured values of R_1 and R_2.

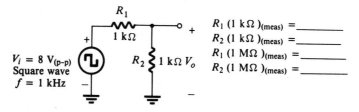

R_1 (1 kΩ)(meas) = _____
R_2 (1 kΩ)(meas) = _____
R_1 (1 MΩ)(meas) = _____
R_2 (1 MΩ)(meas) = _____

Figure 16-4 Loading effects.

Procedure

b. Set V_i to an 8 $V_{p\text{-}p}$ square wave at a frequency of 1 kHz centered on the horizontal center line of the display. Adjust the horizontal sensitivity to show one or two full cycles of V_i.

c. Using the measured resistor values calculate the peak-to-peak values of V_o.

(calculated) $V_{o(p\text{-}p)}$ = _____

d. Energize the network of Fig. 16.4 and measure the output voltage V_o using the oscilloscope.

(measured) $V_{o\,(p\text{-}p)}$ = _____

How do the results of parts 4(c) and 4(d) compare?

e. Now replace the two 1 kΩ resistors with 1 MΩ resistors. Insert the measured values of R_1 and R_2.

f. Using the measured resistor values calculate the peak-to-peak voltage for V_o using the oscilloscope.

(calculated) $V_{o\,(p\text{-}p)}$ = _____

g. Energize the network and measure V_o.

(measured) $V_{o\,(p\text{-}p)}$ = _____

How do the results of Parts 4(f) and 4(g) compare?

h. It is expected that the results of Part 4(**g**) will reveal that the measured and calculated values of V_o do not compare as they did for Parts 4(**c**) and 4(**d**). The change in response is due to the loading of the scope on the circuit when applied to measure V_o. In Fig. 16.5 an additional resistor has been added to that of Fig. 16.4 to represent the loading of the scope on the circuit.

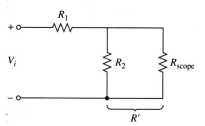

Figure 16-5 Loading of the scope.

Using the measured levels of V_o and V_i the magnitude of R_{scope} can be obtained by solving for R_{scope} in the following equation obtained using the voltage-divider rule.

$$\boxed{R' = \frac{R_2 R_{scope}}{R_2 + R_{scope}} = \frac{R_1}{\frac{V_i}{V_o} - 1}} \quad (16.3)$$

Using the measured levels of V_o and V_i determine R_{scope} using Eq. 16.3 and the measured values of R_1 and R_2.

(calculated) R_{scope} = _____

If the input impedance of the oscilloscope is known compare it to the calculated value.

i. If R_1 is maintained at 1 MΩ and R_2 replaced by a 1 kΩ resistor, use the results of Part 4(**h**) to calculate the expected level of V_o.

(calculated) $V_{o(p\text{-}p)}$ = _____

j. Energize the network and measure the resulting peak-to-peak value of V_o.

(measured) $V_{o(p\text{-}p)}$ = _____

k. How do the results of Parts 4(**i**) and 4(**j**) compare?

Problems and Exercises

1. For the network of Fig. 16.1 is it reasonable to assume the capacitor is simply an "open-circuit" for dc conditions and a "short-circuit" for ac conditions? Do the measured values of the experiment support your conclusions? Why?

2. In general, should a DMM or oscilloscope be used for dc measurements? Why? When is it advantageous to use an oscilloscope?

3. a. What are the relative advantages of using a DMM over an oscilloscope for measuring ac quantities?

b. What are some of the relative advantages of using an oscilloscope over a DMM for measuring ac quantities?

4. A sinusoidal signal occupies 5 horizontal divisions with the horizontal sensitivity set at 0.1 ms/div. What are the period and frequency of the waveform?

(calculated) $T =$ _____

(calculated) $f =$ _____

5. Determine the phase shift between the waveforms of Fig. 16.6. For the phase angle chosen which leads which?

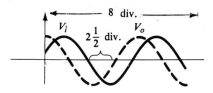

Figure 16-6

(calculated) $\theta =$ _____

6. Derive Eq. 16.3.

Name _____
Date _____
Instructor _____

Common-Emitter Transistor Amplifier

OBJECTIVE

To measure AC and DC voltages in a common-emitter amplifier. To obtain measured values of voltage amplification (A_v), input impedance (Z_i), and output impedance (Z_o), for loaded and unloaded operation.

EQUIPMENT REQUIRED

Instruments

Oscilloscope
DMM
Function Generator
DC power supply

Components

Resistors

(2) 1-kΩ
(2) 3-kΩ
(1) 10-kΩ
(1) 33-kΩ

Capacitors

(2) 15 µF
(1) 100 µF

Transistors

(2) NPN (2N3904, 2N2219, or equivalent general purpose transistor)

EQUIPMENT ISSUED

Item	Laboratory serial no.
DC power supply	
Function generator	
Oscilloscope	
DMM	

RÉSUMÉ OF THEORY

The common-emitter (CE) transistor amplifier configuration is widely used. It provides large voltage gain (typically tens to hundreds) and provides moderate input and output impedance. The AC signal voltage gain is defined as

$$A_v = V_o/V_i$$

where V_o and V_i can both be rms, peak, or peak-peak values. The input impedance, Z_i, is that of the amplifier (as seen by the input signal). The output impedance, Z_o, is that seen looking from the load into the output of the amplifier.

For the voltage-divider DC bias configuration (see Fig. 17.1), all DC bias voltages can be approximately determined without knowing the exact value of transistor beta. The transistor's AC dynamic resistance, r_e, can be calculated using

$$r_e = \frac{26(\text{mV})}{I_{E_Q}(\text{mA})} \quad (17.1)$$

AC Voltage Gain: The AC voltage gain of a CE amplifier (under no-load) can be calculated using

$$A_v = \frac{-R_C}{(R_E + r_e)}$$

If R_E is bypassed by a capacitor use $R_E = 0$ in the above equation.

$$A_v = \frac{-R_C}{r_e} \quad (17.2)$$

AC Input Impedance: The AC input impedance is calculated using

$$Z_i = R_1 || R_2 || \beta(R_E + r_e)$$

If R_E is bypassed by a capacitor use $R_E = 0$ in the above equation.

$$Z_i = R_1 || R_2 || \beta r_e \quad (17.3)$$

AC Output Impedance: The ac output impedance is calculated using

$$Z_o = R_C \quad (17.4)$$

PROCEDURE

Part 1. Common-Emitter DC Bias

a. Insert measured values of each resistor in Fig. 17.1.

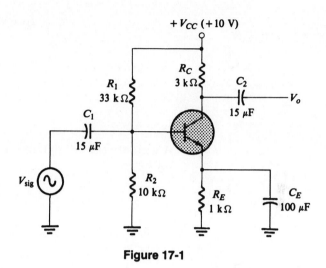

Figure 17-1

b. Calculate DC bias values for the circuit of Fig. 17.1. Record calculated values below.

(calculated) V_B = _____
(calculated) V_E = _____
(calculated) V_C = _____
(calculated) I_E = _____

Calculate r_e using Eq. (17.1) and the calculated level of I_E.

(calculated) r_e = _____

c. Wire up the circuit of Fig. 17.1. Set $V_{CC} = 10$ V. Check the DC bias of the circuit measuring values of

(measured) $V_B = $ _____
(measured) $V_E = $ _____
(measured) $V_C = $ _____

Check that these values compare well with those calculated in step 1b. Calculate the DC emitter current using

$$I_E = V_E/R_E$$

$I_E = $ _____

Calculate the AC dynamic resistance, r_e, using the measured value of I_E.

$$r_e = \frac{26(\text{mV})}{I_E(\text{mA})}$$

$r_e = $ _____

Compare r_e with that calculated in step 1(b).

Part 2. Common-Emitter AC Voltage Gain

a. Calculate the amplifier voltage gain for a bypassed emitter using Eq. (17.2).

(calculated) $A_v = $ _____

b. Apply an AC input signal, $V_{\text{sig}} = 20$ mV, rms at $f = 1$ kHz. Observe the output waveform on the scope to be sure that there is no distortion (if there is, reduce the input signal or check the DC bias). Measure the resulting ac output voltage, V_o, using the scope or a DMM.

(measured) $V_o = $ _____

Calculate the circuit no-load voltage gain using measured values.

$$A_v = \frac{V_o}{V_{sig}}$$

$A_v = $ _____

Compare the measured value of A_v with that calculated in step **a**.

Part 3. AC Input Impedance, Z_i

a. Calculate Z_i using Eq. (17.3). Use beta measured with a transistor curve tracer, beta tester, or the nominal listed value in spec sheets (say, $\beta = 150$).

(calculated) $Z_i = $ _____

b. To measure Z_i connect an input measurement resistor, $R_x = 1\ \text{k}\Omega$, as shown in Fig. 17.2. Apply input $V_{sig} = 20$ mV, rms. Observe the output waveform with a scope to insure that no distortion is present (adjust input amplitude if necessary). Measure V_i.

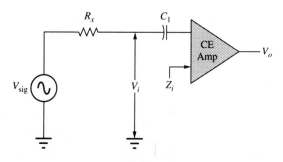

Figure 17-2

Solving for V_i using

(measured) $V_i =$ _____

$$V_i = \frac{V_{sig}}{(Z_i + R_x)} Z_i$$

we get

$$Z_i = \frac{V_i}{(V_{sig} - V_i)} R_x$$

$Z_i =$ _____

Compare the measured value of Z_i with that calculated in step **a**.

Part 4. Output Impedance, Z_o

a. Calculate Z_o using Eq. (17.4).

(calculated) $Z_o =$ _____

b. Remove input measurement resistor, R_x. For input of $V_{sig} = 20$ mV rms, measure the output voltage, V_o. Check output waveform to insure that no distortion is present.

[measured] V_o (unloaded) $= V_o =$ _____

Now connect load $R_L = 3$ kΩ and measure V_o.

[measured] V_o (loaded) $= V_L =$ _____

The output impedance can be obtained from

$$V_L = \frac{R_L}{(Z_o + R_L)} V_o$$

for which

$$Z_o = \frac{V_o - V_L}{V_L} R_L$$

$Z_o =$ _____

Compare the measured value of Z_o with that calculated in step **a**.

Part 5. Oscilloscope Measurement

Connect the amplifier of Fig. 17.1. For input of V_{sig} = 20 mV, p-p, at a frequency of f = 1 kHz, sketch the waveforms for V_{sig} and V_o in Fig. 17.3.

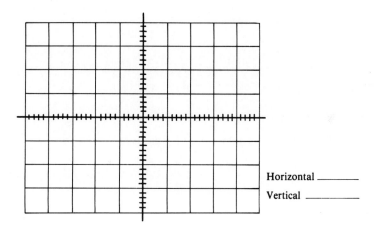

Figure 17-3

Part 6. Computer Analysis

Enter the following PSpice program to analyze the circuit of Fig. 17.1.

```
VCC     5   0   10V
VSIG    1   0   AC    50MV
C1      1   2   2UF
R1      5   2   33K
R2      2   0   10K
RE      3   0   1K
RC      4   5   3K
Q1      4   2   3     QN
.MODEL  QN  NPN (BF=150)
.OP
.AC  LIN  1  1KHZ 1KHZ
.PRINT AC  V(1)  V(2)  V(3)  (4)
.OPTIONS NOPAGE
.END
```

Name _____
Date _____
Instructor _____

EXPERIMENT 18

Common-Base and Emitter-Follower (Common-Collector) Transistor Amplifiers

OBJECTIVE

To measure DC and AC voltages in common-base & emitter-follower (common-collector) amplifiers. To obtain measured values of voltage amplification (A_v), input impedance (Z_i) and output impedance (Z_o).

EQUIPMENT REQUIRED

Instruments

Oscilloscope
DMM
Function Generator
DC power supply

Components

Resistors

(1) 100 Ω
(1) 1-kΩ
(2) 3-kΩ
(2) 10-kΩ
(1) 33-kΩ
(1) 100-kΩ

Capacitors

(2) 15 μF
(1) 100 μF

Transistors

(2) NPN (2N3904, 2N2219, or equivalent general purpose)

EQUIPMENT ISSUED

Item	Laboratory serial no.
DC power supply	
Function generator	
Oscilloscope	
DMM	

RÉSUMÉ OF THEORY

The common-base (CB) transistor amplifier configuration is used primarily for higher frequency operation. It provides large voltage gain at low input and moderate output impedance.

$$A_v = \frac{R_C}{r_e} \tag{18.1}$$

AC Input Impedance: The ac input impedance is calculated using

$$Z_i = r_e \quad \text{(grounded based terminal)} \tag{18.2}$$

AC Output Impedance: The AC output impedance is

$$Z_o = R_C \tag{18.3}$$

The common-collector (CC), or emitter-follower (EF) transistor amplifier configuration is used primarily for impedance matching operation. It provides voltage gain near unity, high input and low output impedance.

AC Voltage Gain: The AC voltage gain of a CC amplifier can be calculated using

$$A_v = \frac{R_E}{R_E + r_e} \tag{18.4}$$

AC Input Impedance: The ac input impedance is

$$Z_i = R_1 || R_2 || \beta(R_E + r_e) \tag{18.5}$$

AC Output Impedance: The AC output impedance is

$$Z_o = r_e \tag{18.6}$$

PROCEDURE

Part 1. Common-Base DC Bias

a. Calculate DC bias values for the circuit of Fig. 18.1. Record calculated values below.

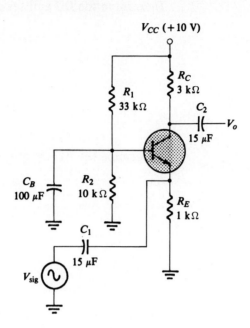

Figure 18-1

(calculated) V_B = _____
(calculated) V_E = _____
(calculated) V_C = _____
(calculated) I_E = _____

Calculate r_e using $r_e = 26(\text{mV})/I_E(\text{mA})$.

(calculated) r_e = _____

Exp. 18 / Common-Base & Emitter-Follower (Common-Collector) Transistor Amplifiers

b. Wire up the circuit of Fig. 18.1. Set $V_{CC} = 10$ V. Check the DC bias of the circuit measuring values of

(measured) $V_B =$ _____
(measured) $V_E =$ _____
(measured) $V_C =$ _____

Determine the DC emitter current using

$$I_E = V_E/R_E$$

$I_E =$ _____

Determine the AC dynamic resistance, r_e

$$r_e = 26(\text{mA})/I_E \text{ (mA)}$$

$r_e =$ _____

Compare the DC voltages, current I_E, and dynamic resistance r_e calculated in step **a** with the values obtained in step **b**.

Part 2. Common-Base AC Voltage Gain

a. Calculate the AC voltage gain of the CB amplifier in Fig. 18.1 using Eq. (18.1).

(calculated) $A_v =$ _____

b. Apply an AC input signal, $V_{sig} = 50$ mV, rms. Measure the resulting AC output voltage, V_o.

(measured) $V_o =$ _____

Determine the circuit AC voltage gain

$$A_v = \frac{V_o}{V_{sig}}$$

Procedure

$A_v =$ _____

Compare the voltage gain calculated in step **a** with that measured in step **b**.

Using the oscilloscope observe and record input waveform, V_{sig} and output waveform, V_o in Fig. 18.2.

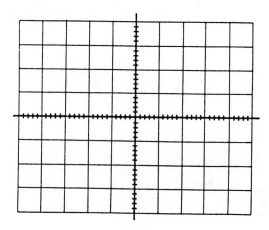

Figure 18-2

Part 3. CB Input Impedance, Z_I

a. Obtain the AC input impedance of the CB amplifier in Fig. 18.1 using Eq. (18.2).

(calculated) $Z_i =$ _____

b. To measure Z_i connect input measurement resistor, $R_x = 100\ \Omega$ as shown in Fig. 18.3. Apply input $V_{sig} = 50$ mV, rms at frequency $f = 1$ kHz. Measure V_i.

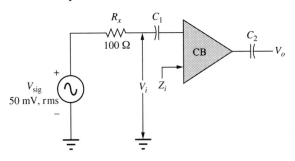

Figure 18-3

(measured) $V_i =$ _____

Determine Z_i using

$$V_i = \frac{Z_i}{(Z_i + R_x)} V_{sig}$$

$$Z_i = \frac{V_i}{(V_{sig} - V_i)} R_x$$

(measured) $Z_i =$ _____

Remove resistor R_x.

Compare the AC input impedance calculated in step **a** with that measured in step **b**.

Part 4. CB Output Impedance, Z_o

a. Determine the AC output impedance of the CB amplifier of Fig. 18.1 using Eq. (18.3).

(calculated) $Z_o =$ _____

Procedure

b. For an input of $V_{sig} = 20$ mV, rms measure the output voltage, V_o when no load is connected.

(measured) V_o (unloaded) = _____

Now connect load $R_L = 3$ kΩ and measure V_L.

(measured) V_L = _____

The output impedance can be determined from

$$V_L = \frac{R_L}{(Z_o + R_L)} V_o$$

for which

$$Z_o = \frac{V_o - V_L}{V_L} R_L$$

Z_o = _____

Compare the AC output impedance calculated in step **a** with the measured in step **b**.

Part 5. Emitter-Follower DC Bias

a. Calculate DC bias values for the EF circuit of Fig. 18.4. Record calculated values below.

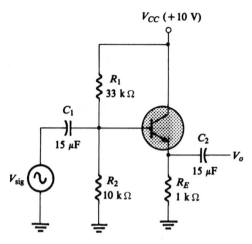

Figure 18-4

(calculated) $V_B =$ _____
(calculated) $V_E =$ _____
(calculated) $V_C =$ _____
(calculated) $I_E =$ _____

Calculate r_e using $r_e = 26(\text{mV})/I_E$ (mA).

(calculated) $r_e =$ _____

b. Wire up the circuit of Fig. 18.4. Set $V_{CC} = 10$ V. Check the DC bias of the circuit measuring values of

(measured) $V_B =$ _____
(measured) $V_E =$ _____
(measured) $V_C =$ _____

Determine I_E using

$$I_E = \frac{V_E}{R_E}$$

$I_E =$ _____

Procedure

Determine the value of r_e using

$$r_e = \frac{26(\text{mV})}{I_E}$$

$r_e = $ _____

Compare the DC voltages and current calculated in step **a** with those measured in step **b**.

Part 6. Emitter-Follower AC Voltage Gain

a. Calculated the AC voltage gain of an EF amplifier using Eq. (18.4).

(calculated) $A_v = $ _____

b. Apply an AC input signal, $V_{sig} = 1$ V, rms. Measure the resulting AC output voltage, V_o.

(measured) $V_o = $ _____

Determine the circuit AC voltage gain

$$A_v = \frac{V_o}{V_{sig}}$$

(measured) $A_v = $ _____

Compare the voltage gain calculated in step **a** with that measured in step **b**.

Observe and record the input signal, V_{sig} and output voltage, V_o in Fig. 18.5.

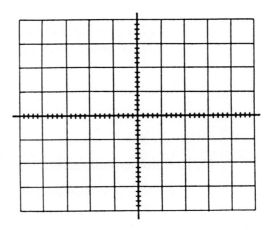

Figure 18-5

Part 7. Emitter Follower (EF) Input Impedance, Z_I

a. Calculate the AC input impedance of an EF amplifier using Eq. (18.5).

(calculated) $Z_i =$ _____

b. To measure Z_i connect input measurement resistor, $R_x = 10$ kΩ as shown in Fig. 18.6. Apply input $V_{sig} = 2$ V, rms at frequency $f = 1$ kHz. Measure V_i.

Procedure

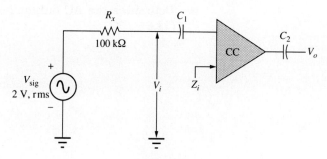

Figure 18-6

(measured) $V_i =$ _____

Calculate Z_i using

$$V_i = \frac{Z_i}{(Z_i + R_x)} V_{sig}$$

$$Z_i = \frac{V_i}{(V_{sig} - V_i)} R_x$$

$Z_i =$ _____

Compare the AC input impedance of a CC amplifier calculated in step **a** with that measured in step **b**.

Part 8. Emitter Follower (EF) Output Impedance, Z_o

a. Determine the AC output impedance of a CC amplifier using Eq. (18.6).

(calculated) $Z_o = $ _____

b. For input of $V_{sig} = 20$ mV, rms at frequency $f = 1$ kHz measure the output voltage, V_o.

(measured) $V_o = $ _____

Now connect load $R_L = 100$ Ω and measure V_L.

(measured) $V_L = $ _____

The output impedance can be calculated from

$$V_L = \frac{R_L}{(Z_o + R_L)} V_o$$

for which

$$Z_o = \frac{V_o - V_L}{V_L} R_L$$

$Z_o = $ _____

Compare the CC output impedance calculated in step **a** with that determined in step **b**.

Part 9. Computer Analysis

Enter the following PSpice program to analyze the circuit of Fig. 18.1.

```
VCC    5    0    10V
VSIG   1    0    AC    1V
C1     1    3    15UF
R1     5    2    33K
R2     2    0    10K
RE     3    0    1K
RC     4    5    3K
Q1     4    2    3     QN
.MODEL QN NPN (BF=100)
.OP
.AC LIN 1 1KHZ 1KHZ
.PRINT AC V(1) V(2) V(3) V(4)
.OPTIONS NOPAGE
.END
```

Enter the following PSpice program to analyze the circuit of Fig. 18.3.

```
VCC    5    0    10V
VSIG   1    0    AC    1V
C1     1    2    15UF
R1     5    2    33K
R2     2    0    10K
RE     3    0    1K
RC     4    5    3K
Q1     4    2    3     QN
.MODEL QN NPN (BF=100)
.OP
.AC LIN 1 1KHZ 1KHZ
.PRINT AC V(1) V(2) V(3) V(4)
.OPTIONS NOPAGE
.END
```

Name _____
Date _____
Instructor _____

EXPERIMENT 19

Design of Common-Emitter Amplifier

OBJECTIVE

To design, build and test a common-emitter amplifier. Both DC bias and AC amplification values are considered.

EQUIPMENT REQUIRED

Instruments

Oscilloscope
DMM
Function Generator
DC power supply

Components

Resistors

To be selected in design

Capacitors

To be selected in design

Transistors

(1) NPN (2N3904, 2N2219, or equivalent general purpose)

EQUIPMENT ISSUED

Item	Laboratory serial no.
DC power supply	
Function generator	
Oscilloscope	
DMM	

RÉSUMÉ OF THEORY

This lab will provide a design for a common-emitter amplifier as shown in Fig. 19.1. The design process begins with a set of specifications that define both the transistor and the circuit operation desired. Fig. 19.1 shows a voltage-divider amplifier with emitter resistor R_E fully bypassed. If possible, a computer should be used to perform the design and test the circuit before it is built. Either PSpice or Microcap II can be used to test any circuit design obtained. Using a 2N3904 (or equivalent transistor), design specifications are:

$\beta = 100$ typical
$I_C(\max) = 200$ mA
$V_{CE}(\max) = 40$ V

The circuit should have the following features:

$A_v = 100$ minimum
$Z_i = 1$ kΩ minimum
$Z_o = 10$ kΩ maximum
$V_{CC} = 10$ V
AC output voltage swing = 3 Vp-p maximum
Load resistance, $R_L = 10$ kΩ minimum

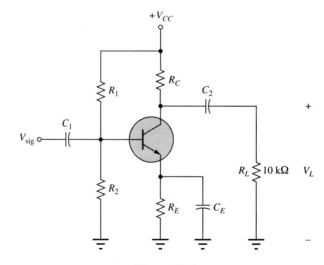

Figure 19-1

Procedure

PROCEDURE

Part 1. Selection of Components

The CE circuit to be built is that shown in Fig. 19.1. The level of V_{CC} (10 V) is well within the maximum rating of the transistor (V_{CE} = 40 V maximum), and would allow a 3 V, p-p output voltage swing. For the mid-band frequency of f = 1 kHz, capacitor values of $C_1 = C_2$ = 15 µF and C_E = 100 µF would be satisfactory.* For the transistor consider a minimum β = 100 in the design.

a. Select

$$V_E = \frac{V_{CC}}{10} = \frac{10 \text{ V}}{10} = 1 \text{ V}.$$

b. Have each squad design for a different value of I_C.** We'll show a design here for I_C = 1 mA. For $I_E = I_C$ = 1 mA, the value of R_E should be

$$R_E = \frac{V_E}{I_E} = \frac{1 \text{ V}}{1 \text{ mA}} = 1 \text{ k}\Omega$$

c. Select R_C to bias circuit at about V_{CE} = 5 V (one-half V_{CC}). Then $V_{R_C} = I_C R_C$ = 4 V, and

$$R_C = \frac{V_{R_C}}{I_C} = \frac{4 \text{ V}}{1 \text{ mA}} = 4 \text{ k}\Omega \text{ (use 4.1 k}\Omega\text{)}$$

d. Check A_v:

$$r_e = \frac{26 \text{ mV}}{I_E \text{ (mA)}} = \frac{26}{1} = 26 \text{ }\Omega$$

$$|A_v| = \frac{R_C}{r_e} = \frac{4.1 \text{ k}\Omega}{26 \text{ }\Omega} = 158$$

e. Since the input impedance looking into the base of the transistor is βr_e = 100 (26 Ω) = 2.6 kΩ, select R_1 and R_2 as large as possible but still sensitive to the dc condition $\beta R_E \geq 10R_2$ so the system is not loaded down.

Using $\beta R_E \geq 10R_2$

we find $R_2 \leq \dfrac{\beta R_E}{10} = \dfrac{(100)(1 \text{ k}\Omega)}{10}$ =10 kΩ

∴ use R_2 = 10 kΩ

---

*For 15 µF: $X_C = 1/(2\pi f C) = 1/[2\pi(1\times 10^3)(15\times 10^{-6})]$ = 10.6 Ω.
For 100 µF: $X_C = 1/(2\pi f C) = 1/[2\pi(1\times 10^3)(100\times 10^{-6})]$ = 1.6 Ω.
**Squad 1 to use I_C = 1 mA, squad 2 to use I_C = 2 mA, etc.

Substituting into the basic equation

$$V_B = \frac{R_2 V_{CC}}{R_1 + R_2} = V_E + 0.7\text{ V} = 1\text{ V} + 0.7\text{ V} = 1.7\text{ V}$$

we have $\dfrac{10\text{ k}\Omega(10\text{ V})}{R_1 + 10\text{ k}\Omega} = 1.7\text{ V}$

and $100\text{ k}\Omega = 1.7\,R_1 + 17\text{ k}\Omega$

or $1.7\,R_1 = 83\text{ k}\Omega$

and $R_1 = \dfrac{83\text{ k}\Omega}{1.7} \cong 48.82\text{ k}\Omega$ (use 47 kΩ)

f. Check Z_i:

$$Z_i = R_1 || R_2 || \beta r_e = 130\text{ k}\Omega || 27\text{ k}\Omega || 100(26\text{ }\Omega) = 2.3\text{ k}\Omega$$

g. Check Z_o:

$$Z_o = R_C = 4.1\text{ k}\Omega$$

Part 2. Computer Analysis of Design

Use a computer analysis to check the design values obtained in Part 1. Either PSpice, Microcap II, or BASIC programs could easily provide this design check. For the circuit of Fig. 19.1, with values obtained in Part 1, the following PSpice analysis could be used.

Check on circuit design in Lab 18 - Fig. 19.1.

```
VCC    5    0    10V
VSIG   1    0    AC      10MV
C1     1    2    15UF
R1     5    2    27K
R2     2    0    5.6K
RC     5    4    4.1K
RE     3    0    1K
CE     3    0    100UF
C2     4    6    15UF
RL     6    0    10K
Q1     4    2    3       QN
.MODEL QN NPN (BF=100)
.OP
.AC  LIN  1  1KHZ  1KHZ
.PRINT AC V(1) V(2) V(4) V(6)
.OPTIONS NOPAGE
.END
```

A PSpice analysis will provide the following results:

```
****      SMALL SIGNAL BIAS SOLUTION
(   1)   0.0000   (  2)   1.6762   (  3)    .9049
(   4)   6.3268   (  5)  10.0000   (  6)   0.0000

VOLTAGE SOURCE CURRENTS
NAME     CURRENT
VCC      -1.204E-03
```

Procedure

```
****        AC ANALYSIS
FREQ         V(1)            V(2)           V(4)           V(6)
1.000E+03    1.000E-02       9.998E-03      1.005E+00      1.005E+00
```

The resulting DC bias values agree well with those desired in the above design. The AC voltage gain is also nearly 100, as desired.

Part 3. Build and Test CE Circuit.

a. Build the CE amplifier circuit of Fig. 19.1 using the capacitors, resistors, and transistor from the design in Part 1, and analysis in Part 2.

b. Set $V_{CC} = 10$ V. Measure and record DC voltages.

(measured) $V_B =$ _____

(measured) $V_E =$ _____

(measured) $V_C =$ _____

Calculate the value of $I_C = I_E$.

$I_C = I_E =$ _____

Calculate dynamic resistance, r_e.

$r_e =$ _____

c. Apply an AC input, $V_{sig} = 10$ mV, rms at $f = 1$ kHz (or adjust value for maximum undistorted load voltage as observed using scope). Measure and record AC voltages.

$V_{sig} =$ _____

(measured) $V_L =$ _____

Calculate A_v with load resistor connected.

$A_v =$ _____

d. Connect measurement resistor $R_x = 3$ kΩ in series with input V_{sig}. Using a DMM measure and record V_{sig} and V_i (from base to ground).

$V_{sig} =$ _____
(measured) $V_i =$ _____

Calculate Z_i.

$Z_i =$ _____

Remove resistor R_x.

e. Remove load resistor R_L. (Readjust V_{sig} if waveform seen using scope is distorted). Measure unloaded ac output voltage.

(measured) $V_o =$ _____

Calculate AC output impedance (using V_L in step **c**).

$Z_o =$ _____

f. Provide a summary of original design specs and actual specs determined by measurement.

Provide a comparison to show whether the design procedure was successful. Indicate any possible factors if design results are not fully satisfactory.

Name _____
Date _____
Instructor _____

EXPERIMENT 20
Common-Source Transistor Amplifier

OBJECTIVE

To measure DC and AC voltages in a common-source amplifier. To obtain measured values of voltage amplification (A_v), input impedance (Z_i) and output impedance (Z_o).

EQUIPMENT REQUIRED

Instruments

Oscilloscope
DMM
Function Generator
DC Supply

Components

Resistors

(1) 510-Ω
(1) 1-kΩ
(1) 2.4-kΩ
(1) 10-kΩ
(2) 1-MΩ

Capacitors

(2) 15-µF
(1) 100-µF

Transistors

(1) 2N3823, or equivalent

EQUIPMENT ISSUED

Item	Laboratory serial no.
DC power supply	
Function generator	
Oscilloscope	
DMM	

RÉSUMÉ OF THEORY

The DC bias of a JFET is determined by the device transfer characteristic (V_P and I_{DSS}) and the DC self-bias determined by the source resistor. The AC voltage gain at this DC bias point is then dependent on the device parameters (g_m or g_{fs}) and circuit drain resistance.

AC Voltage Gain: The voltage gain of the amplifier as in Fig. 20.1 can be calculated from

$$A_v = \frac{V_o}{V_i} = -g_m R_D \qquad [= -g_m(R_D || R_L)] \tag{20.1}$$

where,

$$g_m = g_{m0}(1 - V_{GSQ}/V_P) \quad \text{with } g_{m0} = \frac{2I_{DSS}}{V_P} \tag{20.2}$$

AC Input Impedance: The AC input impedance is

$$Z_i = R_G \tag{20.3}$$

AC Output Impedance: The AC output impedance is

$$Z_o = R_D \tag{20.4}$$

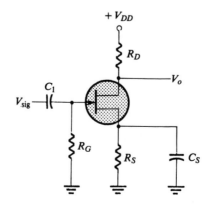

Figure 20-1

Procedure

PROCEDURE

Part 1. Measurement of I_{DSS} and V_P

Use a characteristic curve tracer to determine the values of I_{DSS} and V_P, if available. Otherwise, use the following steps to obtain these values.

    a. Construct the circuit of Fig. 20.1 with V_{DD} = +20 V, R_G = 1 MΩ, R_D = 510 Ω, and R_S = 0 Ω. Measure and record.

(measured) V_D = _____

Calculate the value of drain current, I_D

$$I_D = \frac{V_{DD} - V_D}{R_D}$$

(calculated) I_D = _____

Since this is the drain current at V_{GS} = 0 V

$I_{DSS} = I_D$ = _____

(using the value of I_D just calculated).

    b. Now connect R_S = 1 kΩ. Measure and record the values of

(measured) V_{GS} = _____
(measured) V_D = _____

Using the measured values just obtained:

Calculate V_P as follows.

First: $I_D = \dfrac{V_{DD} - V_D}{R_D}$

(calculated) I_D = _____

Second: $$V_P = \frac{V_{GS}}{1 - \sqrt{\dfrac{I_D}{I_{DSS}}}}$$

(calculated) $V_P =$ _____

Part 2. DC Bias of Common-Source Circuit

a. Calculate the DC bias expected in the circuit of Fig. 20.2, using I_{DSS} and V_P obtained in Part 1.

Draw graphs of the equations

$$I_D = I_{DSS}\left(1 - \frac{V_{GS}}{V_P}\right)^2 \text{ and } V_{GS} = -I_D R_S$$

to graphically obtain the equation intersection.

Or: use a computer or programmable calculator to solve the simultaneous equations.

The calculated DC bias values are:

(calculated) $V_{GS} =$ _____
(calculated) $I_D =$ _____

Using

$$V_D = V_{DD} - I_D R_D$$

(calculated) $V_D =$ _____

Procedure

b. Build the circuit of Fig. 20.2 using R_G = 1MΩ, R_S = 510 Ω, and R_D = 2.4 kΩ. Set V_{DD} = +20 V.

c. Measure the DC bias voltages

(measured) V_G = _____

(measured) V_S = _____

(measured) V_D = _____

(measured) V_{GS} = _____

Determine the value of I_D under DC bias conditions

$$I_D = \frac{V_S}{R_S}$$

I_D = _____

Compare the DC bias values calculated in step **a** with those measured in step **c**.

Part 3. AC Voltage Gain of Common-Source Amplifier

a. Calculate the voltage gain of the common-source amplifier of Fig. 20.2.

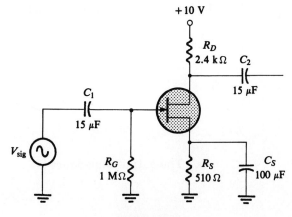

Figure 20-2

$$A_v = -g_m R_D$$

$$\text{with } g_m = \left(\frac{2I_{DSS}}{|V_P|}\right)\left(1 - \frac{V_{GS}}{V_P}\right)$$

Using V_P, I_{DSS} from Part 1, and V_{GS} calculated in Part 2

(calculated) $A_v =$ _____

b. Connect input of $V_{\text{sig}} = 100$ mV at 1 kHz. Measure and record using DMM:

(measured) $V_o =$ _____

Calculate the voltage gain of the amplifier

$$A_v = \frac{V_o}{V_{\text{sig}}}$$

$A_v =$ _____

Part 4. Input and Output Impedance Measurements

a. The input impedance is

$$Z_i = R_G$$

(expected) $Z_i =$ _____

b. The output impedance is

$$Z_o = R_D$$

(expected) $Z_o =$ _____

Procedure

c. Connect a 1 MΩ resistor, R_x in series with the input signal, $V_{sig} = 100$ mV, rms at $f = 100$ Hz. Measure V_i.

(measured) $V_i = $ _____

Determine the input impedance using

$$Z_i = \frac{Z_i}{V_{sig} - V_i} R_x$$

(calculated) $Z_i = $ _____

Remove measurement resistor, R_x.

d. Measure V_o

(measured) $V_o = $ _____

Connect load $R_L = 10$ kΩ. Measure voltage across load, V_L.

(measured) $V_L = $ _____

Determine the AC output impedance using

$$Z_o = \frac{V_o - V_L}{V_L} R_L$$

(calculated) $Z_o = $ _____

Compare the input impedance calculated in step **a** with that determined from measurements in step **c**.

Compare the output impedance calculated in step **b** with that determined from measurements in step **d**.

EXPERIMENT 21

Multistage Amplifier: *RC* Coupling

OBJECTIVE

To measure DC and AC voltages in a multistage FET amplifier; also, to obtain measured values of voltage amplification (A_v), input impedance (Z_i) and output impedance (Z_o).

EQUIPMENT REQUIRED

Instruments

Oscilloscope
DMM
Function Generator
DC Supply

Components

Resistors

(2) 510-Ω
(1) 1-kΩ
(2) 2.4-kΩ
(1) 10-kΩ
(3) 1-MΩ

Capacitors

(3) 15-μF
(2) 100-μF

Transistors

(2) 2N3823, or equivalent

EQUIPMENT ISSUED

Item	Laboratory serial no.
DC Power Supply	
Function Generator	
Oscilloscope	
DMM	

RÉSUMÉ OF THEORY

The DC bias of a JFET is determined by the device transfer characteristic (V_P and I_{DSS}) and the external circuit connected to it. The AC voltage gain at this DC bias point is then dependent on the device parameters (g_m or g_{fs}) and circuit drain resistance.

AC Voltage Gain: The voltage gain of an amplifier stage as in Fig. 21.1 can be calculated from

$$A_v = \frac{V_o}{V_i} = -g_m R_D \; [\; = -g_m(R_D||R_L)] \tag{21.1}$$

where

$$g_m = g_{m0}\left(1 - \frac{V_{GSQ}}{V_P}\right) \quad \text{with } g_{m0} = \frac{2I_{DSS}}{|V_p|} \tag{21.2}$$

AC Input Impedance: The AC input impedance is

$$Z_i = R_G \tag{21.3}$$

AC Output Impedance: The AC output impedance is

$$Z_o = R_D \tag{21.4}$$

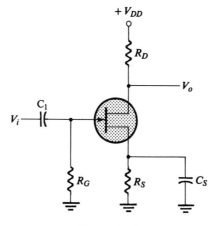

Figure 21-1

Procedure

PROCEDURE

Part 1. Measurement of I_{DSS} and V_P

It is necessary to obtain values of I_{DSS} and V_P for both Q_1 and Q_2. Use a characteristic curve tracer, if available, to determine the values of I_{DSS} and V_P. Obtain readings at $V_{DS} = +10$ V.

For Q_1:

$I_{DSS} =$ _____
$V_P =$ _____

For Q_2:

$I_{DSS} =$ _____
$V_P =$ _____

Go on to Part 2.

Otherwise, use the following steps to obtain these values.

a. Construct the circuit of Fig. 21.2 with $R_D = 510$ Ω but with $R_S = 0$ Ω. Measure and record.

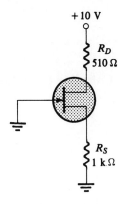

Figure 21-2

(measured) $V_D =$ _____

Calculate the value of drain current, I_D

$$I_D = \frac{V_{DD} - V_D}{R_D}$$

(calculated) I_D = _____

since this is the drain current at $V_{GS} = 0$ V

$I_{DSS}(Q_1) = I_D$ = _____

(using the value of I_D just calculated).

Replace Q_1 and repeat measurement with Q_2.

(measured) V_D = _____

Calculate the value of drain current, I_D

$$I_D = \frac{V_{DD} - V_D}{R_D}$$

(calculated) I_D = _____

since this is the drain current at $V_{GS} = 0$ V

$I_{DSS}(Q_2) = I_D$ = _____

(using the value of I_D just calculated).

Procedure

b. Now connect $R_S = 1$ kΩ. Measure and record the values of

(measured) $V_{GS} =$ _____
(measured) $V_D =$ _____

Using the measured values just obtained, calculate V_P as follows.

$$I_D = \frac{V_{DD} - V_D}{R_D}$$

(calculated) $I_D =$ _____

$$V_P = \frac{V_{GS}}{1 - \sqrt{\dfrac{I_D}{I_{DSS}}}}$$

(calculated) $V_P(Q_2) =$ _____

Replace transistor Q_2 and repeat step **b** measurements.

(measured) $V_{GS} =$ _____
(measured) $V_D =$ _____

Using the measured values just obtained, calculate V_P as follows.

$$I_D = \frac{V_{DD} - V_D}{R_D}$$

(calculated) $I_D =$ _____

$$V_P = \frac{V_{GS}}{1 - \sqrt{\frac{I_D}{I_{DSS}}}}$$

(calculated) $V_P(Q_1) =$ _____

Part 2. DC Bias of Common-Source Circuit

a. Calculate the DC bias expected in the circuit of Fig. 21.3, using I_{DSS} and V_P obtained in Part 1 for each transistor.

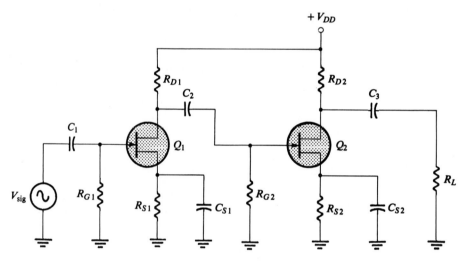

Figure 21-3

Draw graphs of the equations

$$I_D = I_{DSS}\left(1 - \frac{V_{GS}}{V_P}\right)^2 \text{ and } V_{GS} = -I_D R_S$$

to graphically obtain the equation intersection.
Or: Use a computer or programmable calculator to solve the simultaneous equations.
The calculated DC bias values are:

(calculated) $V_{GS1} =$ _____

Procedure

(calculated) I_{D_1} = _____

using

$$V_{D1} = V_{DD} - I_{D_1}R_{D_1}$$

(calculated) V_{D_1} = _____

The calculated DC bias values are:

(calculated) V_{GS_2} = _____
(calculated) I_{D_2} = _____

using

$$V_{D_2} = V_{DD} - I_{D_2}R_{D_2}$$

(calculated) V_{D_2} = _____

b. Build the circuit of Fig. 21.3 using $R_{G_1} = R_{G_2}$ =1MΩ, $R_{S_1} = R_{S_2} =$ 510 Ω, and $R_{D_1} = R_{D_2}$ = 2.4 kΩ. Set V_{DD} = +20 V.

c. Measure the DC bias voltages

(measured) V_{G_1} = _____
(measured) V_{S_1} = _____
(measured) V_{D_1} = _____
(measured) V_{GS_1} = _____

Determine the value of I_D under DC bias conditions (using nominal resistor values)

$$I_{D_1} = \frac{V_{S_2}}{R_{S_2}}$$

I_{D_1} = _____
(measured) V_{G_2} = _____
(measured) V_{S_2} = _____
(measured) V_{D_2} = _____
(measured) V_{GS_2} = _____

Determine the value of I_{D_2} under DC bias conditions

$$I_{D_2} = \frac{V_{S1}}{R_{S1}}$$

$I_{D_2} =$ _____

Compare the DC bias values calculated in step **a** with those measured in step **c**.

Part 3. AC Voltage Gain of Amplifier

a. Calculate the voltage gain of the common-source amplifier of Fig. 21.3.

For stage 2:

$$A_{v_2} = -g_m(R_{D2}||R_L)$$

$$\text{with } g_m(Q_2) = \frac{2I_{DSS}(Q_2)}{|V_p(Q_2)|}\left(1 - \frac{V_{GS_2}}{V_p(Q_2)}\right)$$

Using $V_P(Q_2)$, $I_{DSS}(Q_2)$ from Part 1, and V_{GS_2} calculated in Part 2

(calculated) $A_{V_2} =$ _____

For stage 1:

$$A_{v_1} = -g_{m_1}(R_{D1}||Z_{i_2})$$

$$\text{with } g_m(Q_1) = \frac{2I_{DSS}(Q_1)}{|V_p(Q_1)|}\left(1 - \frac{V_{GS_1}}{V_p(Q_1)}\right)$$

Using $V_P(Q_1)$, $I_{DSS}(Q_1)$ from Part 1, and V_{GS_1} calculated in Part 2

(calculated) $A_{V_1} =$ _____

Calculate overall amplifier gain:

$$A_v = A_{v_1} \times A_{v_2}$$

(calculated) $A_v =$ _____

b. Connect input of $V_{sig} = 10$ mV, rms at $f = 1$ kHz. Use the oscilloscope to obtain an undistorted output voltage, adjusting V_{sig} if necessary. Measure and record:

(measured) $V_{sig} =$ _____

(measured) $V_L =$ _____

Calculate the voltage gain of the overall amplifier

$$A_v = \frac{V_L}{V_{sig}}$$

$A_v =$ _____

Measure and record:

(measured) $V_{o_1} =$ _____

Calculate the gain of each stage:

$$A_{v_1} = \frac{V_{o_1}}{V_{sig}}$$

(measured) $A_{v_1} =$ _____

$$A_{v_2} = \frac{V_L}{V_{o1}}$$

(measured) A_{v_2} = _____

Part 4. Input and Output Impedance Measurements

a. The input impedance is

$$Z_i = R_{G_1}$$

Z_i = _____

b. The output impedance is

$$Z_o = R_{D_2}$$

Z_o = _____

c. Connect a 1 MΩ resistor, R_x in series with the input signal, V_{sig} = 10 mV, rms at f = 100 Hz. Measure V_{i_1}.

(measured) V_{i_1} = _____

Determine the input impedance using

$$Z_i = \frac{V_{i_1}}{V_{sig} - V_{i_1}} R_x$$

Z_i = _____

Remove measurement resistor, R_x.

Procedure

d. Measure V_L

(measured) $V_L =$ _____

Disconnect load $R_L = 10$ kΩ. Measure output voltage, V_o.

(measured) $V_o =$ _____

Determine the AC output impedance using

$$Z_o = \frac{V_o - V_{i_1}}{V_L} R_L$$

$Z_o =$ _____

Compare the input impedance calculated in step **a** with that determined from measurements in step **c**.

Compare the output impedance calculated in step **b** with that determined from measurements in step **d**.

Name _____
Date _____
Instructor _____

CMOS Circuits

OBJECTIVE

To measure DC and AC operation in CMOS circuits.

EQUIPMENT REQUIRED

Instruments

Oscilloscope
DMM
Function Generator
DC Supply

Components

Integrated circuits

(1) 74HC02, or 14002
(1) 74HC04, or 14004

EQUIPMENT ISSUED

Item	Laboratory serial no.
DC Power Supply	
Function Generator	
Oscilloscope	
DMM	

RÉSUMÉ OF THEORY

A CMOS circuit can be built using two opposite type MOSFET devices as shown in Fig. 22.1. Digital inputs are 0 V and 5 V. For an input of 0 V the n-type enhancement MOSFET (nMOS) device is turned *off*, while the p-type enhancement MOSFET (pMOS) device is turned *on* as shown in Fig. 22.2a. An input of 5 V will drive the pMOS device *off* and nMOS device *on* with the output then near 0 V, as shown in Fig. 22.2b.

A CMOS gate with two inputs is shown in Fig. 22.3. Each input is connected to a pair of pMOS and nMOS transistors. The operation for various inputs of 0 V and 5 V is summarized in Figure 22.3. When both inputs are 0 V the two pMOS devices are *on*, both nMOS devices are *off*, and the output is 5 V. When both inputs are 0 V, or even one input 0 V, one nMOS device is *on*, one pMOS device is *off*, and the output is near 0 V.

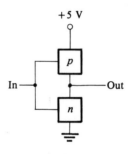

Figure 22-1

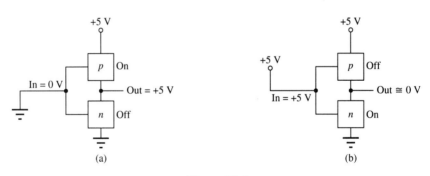

Figure 22-2

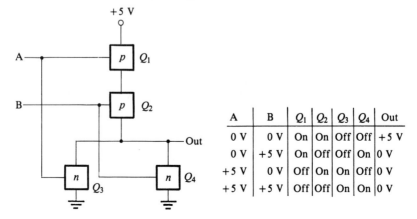

A	B	Q_1	Q_2	Q_3	Q_4	Out
0 V	0 V	On	On	Off	Off	+5 V
0 V	+5 V	On	Off	Off	On	0 V
+5 V	0 V	Off	On	On	Off	0 V
+5 V	+5 V	Off	Off	On	On	0 V

Figure 22-3

PROCEDURE

Part 1. CMOS Inverter Circuit

a. Construct the CMOS inverter circuit shown in Fig. 22.1.

b. For the CMOS inverter circuit of Fig. 22.1 determine the output voltage for inputs of 0 V and 5 V and record in Table 22.1.

TABLE 22.1

IN	OUT
0 V	
5 V	

c. Connect 5 V to an inverter IC such as the 74HC04 or 14004. Apply inputs of 0 V and 5 V, recording outputs in Table 22.2.

TABLE 22.2

IN	OUT
0 V	
5 V	

d. Apply a clock signal ($f = 10$ kHz) as input and record input and output waveforms viewed on oscilloscope in Fig. 22.4.

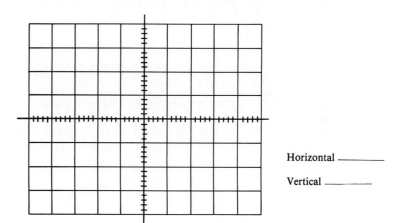

Horizontal _____

Vertical _____

Figure 22-4

Part 2. CMOS Gate

a. Connect power to a CMOS IC such as a 74HC02 or 14002 as shown in Fig. 22.5. Apply inputs of 0 V and 5 V and record output in Table 22.3.

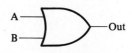

Figure 22-5

TABLE 22.3

A	B	OUTPUT
0 V	0 V	
0 V	5 V	
5 V	0 V	
5 V	5 V	

b. Connect 0 V to one input and a digital clock to the other. Observe and record output waveform in Fig. 22.6a.

c. Connect 5 V to one input and a digital clock to the other. Observe and record output waveform in Fig. 22.6b.

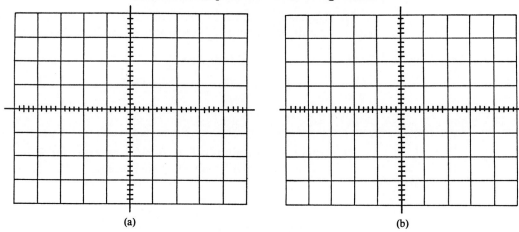

(a)   (b)

Figure 22-6

Part 3. CMOS Input-Output Characteristic

a. Using the CMOS inverter circuit (74HC040 with variable input as shown in Fig. 22.7 complete Table 22.4.

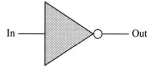

Figure 22-7

TABLE 22.4

IN	0.0	0.2	0.4	0.6	0.8	1.0	1.2	1.4	1.6	1.8	2.0	2.2
OUT												

IN	2.4	2.6	2.8	3.0	3.2	3.4	3.6	3.8	4.0	4.2	4.4	4.6	4.8	5.0
OUT														

Procedure

b. Plot the data from Table 22.4 in Fig. 22.8.

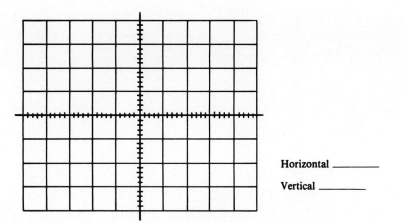

Horizontal _____
Vertical _____

Figure 22-8

Name _____
Date _____
Instructor _____

EXPERIMENT 23

Darlington and Cascode Amplifier Circuits

OBJECTIVE

To measure DC and AC voltages in Darlington and cascode connection circuits.

EQUIPMENT REQUIRED

Instruments

Oscilloscope
DMM
Function Generator
DC Supply

Components

Resistors

(1) 100-Ω
(1) 51-Ω, 1W
(1) 1-kΩ
(1) 1.8-kΩ
(1) 4.7-kΩ
(1) 5.6-kΩ
(1) 6.8-kΩ
(1) 50-kΩ pot
(1) 100-kΩ

Capacitors

(1) 0.001 µF
(4) 10 µF

Transistors

(2) 2N3904, (or equivalent general purpose npn)
(1) TIP120 (npn Darlington)

EQUIPMENT ISSUED

Item	Laboratory serial no.
DC Power Supply	
Function Generator	
Oscilloscope	
DMM	

RÉSUMÉ OF THEORY

Darlington Circuit: A Darlington connection (as shown in Fig. 23.1) provides a pair of BJT transistors in a single IC package with effective beta (β_D) equal to the product of the individual transistor betas.

$$\beta_D = \beta_1 \beta_2 \tag{23.1}$$

The Darlington emitter-follower has a higher input impedance than that of an emitter-follower. The Darlington emitter-follower input impedance is

$$Z_i = R_B || (\beta_D R_E) \tag{23.2}$$

The output impedance of the Darlington emitter-follower is

$$Z_o = r_e \tag{23.3}$$

The voltage gain of a Darlington emitter-follower circuit is

$$A_v = \frac{R_E}{(R_E + r_e)} \tag{23.4}$$

Cascode Circuit: A cascode circuit, as shown in Fig. 23.2, provides a common-emitter amplifier using Q_1 directly connected to a common-base amplifier using Q_2. The voltage gain of stage Q_1 is approximately 1, with the voltage V_{o1} being opposite in polarity to that applied as V_i.

$$A_{v_1} = -1 \tag{23.5}$$

The voltage gain of stage Q_2 is noninverted and of magnitude

$$A_{v_2} = \frac{R_C}{r_{e_2}} \tag{23.6}$$

resulting in an overall gain

$$A_v = A_{v_1} A_{v_2} = -R_C r_{e_2} \tag{23.7}$$

PROCEDURE

Part 1. Darlington Emitter-Follower Circuit

a. For the circuit of Fig. 23.1 calculate the DC bias voltages and currents.

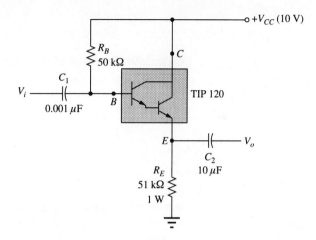

Figure 23-1

(calculated) $V_B =$ _____

(calculated) $V_E =$ _____

Calculate the theoretical values of voltage gain, input and output impedance.

(calculated) $A_V =$ _____

(calculated) $Z_i =$ _____

(calculated) $Z_o =$ _____

b. Construct the Darlington circuit of Fig. 23.1. Adjust the 50-kΩ potentiometer (R_B) to provide an emitter voltage, $V_E = 5$ V. Using a DMM, measure and record the DC bias values:

(measured) $V_B =$ _____

(measured) $V_E =$ _____

Calculate the base and emitter DC currents

(calculated) $I_B =$ _____

(calculated) $I_E =$ _____

Calculate the value of transistor beta at this Q-point:

(calculated) $\beta_D =$ _____

c. Apply an input signal $V_{sig} = 1$ V, peak at $f = 10$ kHz. Using the oscilloscope observe and record the output voltage to assure that the signal is not clipped or distorted. (Reduce the input signal amplitude if necessary.)

(measured) $V_i =$ _____

(measured) $V_o =$ _____

Calculate and record the AC voltage gain

$A_v = V_o/V_i =$ _____

Part 2. Darlington Input and Output Impedance

a. Calculate the input impedance

(calculated) $Z_i =$ _____

Calculate the circuit output impedance

(calculated) $Z_o =$ _____

Procedure

b. Connect a measurement resistor, $R_x = 100$ kΩ, in series with V_{sig}. Measure and record input voltage, V_i.

(measured) $V_i =$ _____

Calculate the circuit input impedance using

$$Z_i = \frac{V_i}{V_{sig} + V_i} R_x$$

(calculated) $Z_i =$ _____

Remove measurement resistor, R_x.

c. Measure the output voltage, V_o with no load connected.

(measured) $V_o =$ _____

Connect load resistor, $R_L = 100$ Ω. Measure and record resulting output voltage:

(measured) $V_o = V_L =$ _____

Calculate the output impedance using

$$Z_o = \frac{V_o - V_L}{V_L} R_L$$

(calculated) $Z_o =$ _____

Compare the calculated and measured values of Z_i and Z_o.

Part 3. Cascode Amplifier

a. Calculate DC bias voltages and currents in the cascode amplifier of Fig. 23.2 (assuming base currents are much less than the voltage divider current).

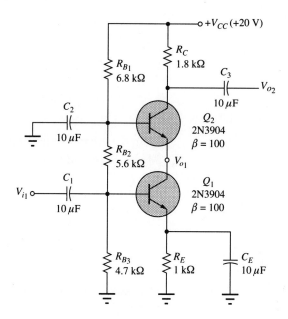

Figure 23-2

(calculated) V_{B_1} = _____
(calculated) V_{E_1} = _____
(calculated) V_{C_1} = _____
(calculated) V_{B_2} = _____
(calculated) V_{E_2} = _____
(calculated) V_{C_2} = _____

Procedure

Calculate the DC bias emitter currents

(calculated) I_{E_1} = _____
(calculated) I_{E_2} = _____

Calculate the transistor dynamic resistances

(calculated) r_{e_1} = _____
(calculated) r_{e_2} = _____

b. Connect the cascode circuit of Fig. 23.2. Measure and record DC bias voltages.

(measured) V_{B_1} = _____
(measured) V_{E_1} = _____
(measured) V_{C_1} = _____
(measured) V_{B_2} = _____
(measured) V_{E_2} = _____
(measured) V_{C_2} = _____

Calculate the values of emitter current

I_{E_1} = _____
I_{E_2} = _____

and the values of dynamic resistance

r_{e_1} = _____
r_{e_2} = _____

c. Using Eqs. (23.5) and (23.6) calculate the AC voltage gain of each transistor stage:

(calculated) A_{v_1} = _____
(calculated) A_{v_2} = _____

d. Apply input signal, V_{sig} = 10 mV, peak at f = 10 kHz. Using the oscilloscope observe the output waveform V_o to make sure that no signal distortion occurs. If the output is clipped or distorted reduce the input signal until the clipping or distortion disappears.

Using the DMM measure, record the AC signals.

(measured) V_i = _____
(measured) V_{o_1} = _____
(measured) V_{o_2} = _____

Calculate the measured voltage gains.

$A_{v_1} = V_{o_1}/V_i$ = _____
$A_{v2} = V_{o_2}/V_{o_1}$ = _____
$A_v = V_{o_2}/V_i$ = _____

Compare the measured voltage gains with those calculated in steps **c** and **d**.

e. Using the oscilloscope, observe and record waveforms for the input signal, V_i, output of stage 1, V_{o_1}, and output of stage 2, V_{o_2}. Show amplitude and phase relations clearly.

Name _____
Date _____
Instructor _____

EXPERIMENT 24

Current Source and Current Mirror Circuits

OBJECTIVE

To measure DC voltages in current source and current mirror circuits.

EQUIPMENT REQUIRED

Instruments

Oscilloscope
DMM
Function Generator
DC Supply

Components

Resistors

(1) 20-Ω
(1) 51-Ω,
(1) 82-Ω
(1) 100-Ω
(1) 150-Ω
(2) 1.2-kΩ
(1) 4.3-kΩ
(1) 5.1-kΩ
(1) 7.5-kΩ
(1) 10-kΩ

Transistors

(3) 2N3904, or equivalent npn transistor
(1) 2N3823, or equivalent JFET n-channel transistor

EQUIPMENT ISSUED

Item	Laboratory serial no.
DC Power Supply	
Function generator	
Oscilloscope	
DMM	

RÉSUMÉ OF THEORY

Current source and current mirror circuits are part of many types of linear integrated circuits. This experiment will provide building and testing a few types of each circuit.

Current Source: Fig. 24.1 shows a simple form of current source using a JFET biased to operate at its drain-source saturation current. Regardless of the load R_L (within practical limits) the current through load R_L will be set by the JFET device:

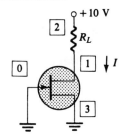

Figure 24-1

$$I_L = I_{DSS} \tag{24.1}$$

A BJT current source circuit is shown in Fig. 24.2. The base voltage is approximately set by

$$V_B = \frac{R_1}{R_1 + R_2}(-V_{EE})$$

The emitter voltage is then

$$V_E = V_B - 0.7 \text{ V}$$

with the emitter current then

$$I_E = \frac{V_E - V_{EE}}{R_E} = I_L \tag{24.2}$$

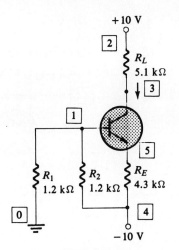

Figure 24-2

Current Mirror: The circuit of a Fig. 24.3 is a current mirror, in which the current set through resistor R_x is mirrored through the load

$$I_x = \frac{V_{CC} - V_{BE}}{R_x} = I_L \qquad (24.3)$$

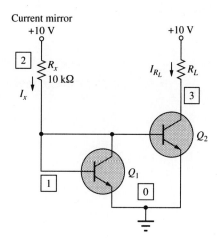

Figure 24-3

The circuit of Fig. 24.4 shows how a current mirror can provide the same current to a number of loads. The mirrored current set through resistor R_x and mirrored through both loads is

$$I_x = \frac{V_{CC} - V_{BE}}{R_X} = I_{L_1} = I_{L_2} \qquad (24.4)$$

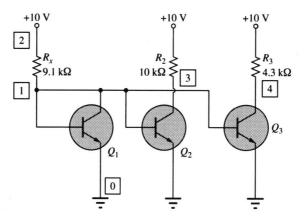

Figure 24-4

PROCEDURE

Part 1. JFET Current Source

a. Wire up the circuit of Fig. 24.1. Use $R_L = 51$. Measure and record the drain-source voltage.

(measured) V_{DS} = _____

b. Using the voltage measured in step **a**, calculate the load current.

$$I_L = \frac{V_{DD} - V_{DS}}{R_L}$$

I_L = _____

c. Replace R_L with resistors as listed in Table 24.1, and repeat steps **a** and **b**.

TABLE 24.1

R_L	20 Ω	51 Ω	82 Ω	100 Ω	150 Ω
V_{DS}					
I_L					

Procedure

Part 2. BJT Current Source

a. Calculate the current through the load in the circuit of Fig. 24.2.

(calculated) $I_L =$ _____

b. Wire up the circuit of Fig. 24.2. Measure and record the following voltages.

(measured) $V_E =$ _____
(measured) $V_C =$ _____

c. Calculate the emitter current and that through the load.

$I_E =$ _____
$I_L =$ _____

d. Replace R_L with resistors listed in Table 24.2 and repeat steps **a** through **c**.

TABLE 24.2

R_L	3.6 kΩ	4.3 kΩ	5.1 kΩ	7.5 kΩ
V_E				
V_C				
I_E				
I_L				

Part 3. Current Mirror

a. Calculate the mirror current in the circuit of Fig. 24.3.

(calculated) $I_x =$ _____

b. Wire up the circuit of Fig. 24.3 and measure

(measured) V_{B_1} = _____
(measured) V_{C_2} = _____
I_x = _____
I_L = _____

c. Change R_L to 3.6 kΩ and repeat steps **a** and **b**.

(calculated) I_x = _____
(measured) V_{B_1} = _____
(measured) V_{C_2} = _____
I_x = _____
I_L = _____

Part 4. Multiple Current Mirrors

a. Calculate the mirror current in the circuit of Fig. 24.4.

(calculated) I_x = _____

b. Wire up the circuit of Fig. 24.4 and measure

(measured) V_{B_1} = _____
(measured) V_{C_2} = _____
(measured) V_{C_3} = _____
I_x = _____
I_{L_1} = _____
I_{L_2} = _____

Procedure

c. Change R_L to 3.6 kΩ and repeat steps **a** and **b**.

(calculated) $I_x =$ _____
(measured) $V_{B_1} =$ _____
(measured) $V_{C_2} =$ _____
(measured) $V_{C_3} =$ _____
$I_x =$ _____
$I_{L_1} =$ _____
$I_{L_2} =$ _____

Name _____
Date _____
Instructor _____

EXPERIMENT 25

Frequency Response of a Common-Emitter Amplifier

OBJECTIVE

To examine the frequency response of a common-emitter amplifier circuit..

EQUIPMENT REQUIRED

Instruments

Oscilloscope
DMM
Function Generator
DC Supply

Components

Resistors

(2) 2.2-kΩ
(1) 3.9-kΩ
(1) 10-kΩ
(1) 39-kΩ

Transistors

(1) 2N3904, (or equivalent general purpose npn)

Capacitors

(1) 1 µF
(1) 10 µF
(1) 20 µF

EQUIPMENT ISSUED

Item	Laboratory serial no.
DC Power Supply	
Function Generator	
Oscilloscope	
DMM	

RÉSUMÉ OF THEORY

The analysis of the frequency response of an amplifier can be considered in three frequency ranges: the low-, mid-, and high-frequency regions. In the low-frequency region the capacitors used for DC isolation (AC coupling) and bypass operation affect the lower cutoff (lower 3 dB) frequency. In the mid-frequency range only resistive elements affect the gain, the gain remaining constant. In the high-frequency region of operation, stray wiring capacitances and device inter-terminal capacitances will determine the circuit's upper cutoff frequency.

Lower Cutoff (lower 3 dB) Frequency: Each capacitor used will result in a cutoff frequency. The lower cutoff frequency is then the largest of these lower cutoff frequencies. For the network of Fig. 25.1 the lower cutoff frequencies are as follows.

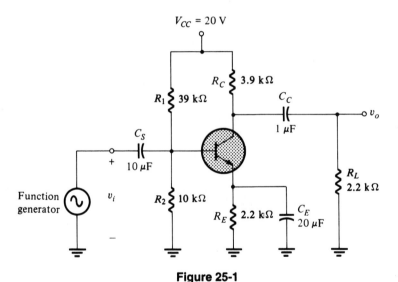

Figure 25-1

C_1: The cutoff frequency due to the input coupling capacitor is

$$f_{L,1} = \frac{1}{2\pi R_i C_i} \text{ Hz} \qquad \text{with: } R_i = R_1 || R_2 || \beta r_e \qquad (25.1)$$

C_2: The cutoff frequency due to the output coupling capacitor is

$$f_{L,2} = \frac{1}{2\pi (R_C + R_L) C_i} \text{ Hz} \qquad (25.2)$$

C_E: The cutoff frequency due to the emitter bypass capacitor is

$$f_{L,E} = \frac{1}{2\pi R_e C_e} \text{ Hz} \qquad \text{with } R_e = R_E || r_e \qquad (25.3)$$

Upper Cutoff (upper 3-dB) Frequency: In the high frequency range the amplifier gain is affected by the transistor's parasitic capacitances as follows:

At input connection of circuit:

$$f_{H,i} = \frac{1}{2\pi R_{Th,i} C_i} \text{ Hz} \qquad (25.4)$$

where

$$R_{Th,i} = R_1 || R_2 || \beta r_e$$

and C_i is

$$C_i = C_{w,i} + C_{be} + (1 + |A_v|)C_{bc}$$

$C_{w,i}$ = input wiring capacitance
A_v = voltage gain of amplifier at mid-band frequency
C_{be} = capacitance between transistor base-emitter terminals
C_{bc} = capacitance between transistor base-collector terminals

At output connection of circuit:

$$f_{H,o} = \frac{1}{2\pi R_{Th,o} C_o} \text{ Hz}$$

where

$$R_{Th,o} = R_C || R_L$$

and

$$C_o = C_{w,o} + C_{ce}$$

$C_{w,o}$ = output wiring capacitance
C_{ce} = capacitance between transistor collector-emitter terminals

(We'll ignore the transistor's upper cutoff frequency, as it usually is greater than that due to wiring and inter-terminal capacitances.)

Keep in mind that the 3-dB cutoff frequencies are defined by 70.7% of the midband gain, or $0.707 A_{v,\text{mid}}$. That is, once the midband gain is measured, the upper and lower cutoff frequencies are measured at the points at which the gain drops to 0.707 the midband gain at either upper or lower frequency.

PROCEDURE

Part 1. Low-Frequency Response Calculations

a. Using the specifications data for the transistor record values

(specified) C_{be} = _____
(specified) C_{bc} = _____
(specified) C_{ce} = _____

Enter values of typical wiring capacitance

(approximated) $C_{w,i}$ = _____
(approximated) $C_{w,o}$ = _____

b. Using a characteristic curve tracer, beta measuring instrument, or value obtained from previous use in the lab, obtain the value of transistor beta.

(measured) β = _____

c. Calculate values of DC bias voltage and current for the circuit of Fig. 25.1.

(calculated) V_B = _____
(calculated) V_E = _____
(calculated) V_C = _____
(calculated) I_E = _____

Using the value of I_E calculate the transistor dynamic resistance.

(calculated) r_e = _____

Procedure

d. Calculate the magnitude of amplifier midband gain (under load) using

$$A_{v,\text{mid}} = \frac{R_C \| R_L}{r_e}$$

e. Calculate lower cutoff frequencies due to coupling capacitors and due to bypass capacity.

(calculated) $f_{L,1}$ = _____
(calculated) $f_{L,2}$ = _____
(calculated) $f_{L,E}$ = _____

Part 2. Low Frequency Response Measurements

a. Construct the network of Fig. 25.1. Record actual resistor values in space provided in Fig. 25.1, if desired. Adjust $V_{CC} = 20$ V. Apply an input AC signal, $V_{\text{sig}} = 20$ mV, peak at a frequency of $f = 5$ kHz. Observe the output voltage using a scope. If V_o shows distortion, reduce V_{sig} until the output is undistorted.

b. Measure and record signals for undistorted operation.

(measured) V_{sig} = _____
(measured) V_o = _____

Calculate the circuit's mid-frequency voltage gain.

$A_{v,\text{mid}}$ = _____

Maintaining the input voltage at the level set above, vary the frequency and measure and record V_o to complete Table 25.1.

TABLE 25.1

f	50-Hz	100-Hz	200-Hz	400-Hz	600-Hz	800-Hz	1-kHz	2-kHz
V_o								

f	3-kHz	5-kHz	10-kHz
V_o			

Calculate the amplifier voltage gain for each frequency and complete Table 25.2.

TABLE 25.2

f	50-Hz	100-Hz	200-Hz	400-Hz	600-Hz	800-Hz	1-kHz	2-kHz
A_v								

f	3-kHz	5-kHz	10-kHz
A_v			

Part 3. High Frequency Response Calculations

a. Using the equations provided in the Resume of Theory calculate upper cutoff frequencies and record below.

(calculated) $f_{H,i}$ = _____

(calculated) $f_{H,o}$ = _____

b. Applying an input which provides non-distorted output voltage complete Table 25.3 measuring the resulting output voltage over a range of high frequency values.

(measured) V_i = _____

Procedure

TABLE 25.3

f	10-kHz	50-kHz	100-kHz	300-kHz	500-kHz	600-kHz	700-kHz
V_o							

f	900-kHz	1-MHz	2-MHz
V_o			

Calculate the amplifier voltage gain (in dB units) and complete Table 25.4.

TABLE 25.3

f	10-kHz	50-kHz	100-kHz	300-kHz	500-kHz	600-kHz	700-kHz
A_V							

f	900-kHz	1-MHz	2-MHz
A_v			

Part 4. Multiple Current Mirrors

a. Using the semi-log paper of Fig. 25.2, plot the gain versus frequency over the full frequency range. Plot the actual points and connect to obtain the actual plot. Use straight-line approximation curves to obtain the Bode plot.

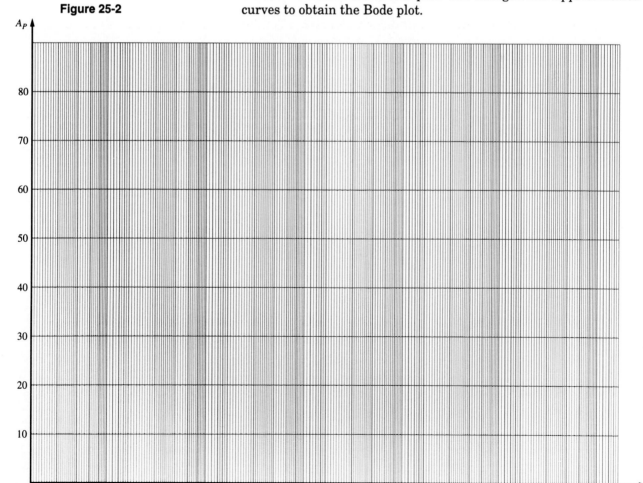

Figure 25-2

b. From the plot obtain the lower and upper 3-dB frequency points and record below.

(measured) f_{-3dB} = _____

(measured) f_{+3dB} = _____

Compare the measured values with those calculated in Parts **1** and **3**.

EXPERIMENT 26

Class-A and Class-B Power Amplifiers

OBJECTIVE

To measure DC and AC voltages, and power input and output for both class-A and class-B power amplifiers.

EQUIPMENT REQUIRED

Instruments

Oscilloscope
DMM
Function Generator
DC power supply

Components

Resistors

(1) 20-Ω
(1) 120-Ω, 0.5-W
(1) 180-Ω
(2) 1-kΩ, 0.5-W
(1) 10-kΩ

Capacitors

(3) 10-µF
(1) 100-µF

Transistors

(1) npn medium power, 15-W (2N4300 or eqv't)

(1) pnp medium power, 15-W (2N5333 or eqv't)
(2) Silicon diode

EQUIPMENT ISSUED

Item	Laboratory serial no.
DC power supply	
Function generator	
Oscilloscope	
DMM	

RÉSUMÉ OF THEORY

A class-A amplifier draws the same power from the voltage supply regardless of the signal applied. The input power is calculated from

$$P_i \text{(DC)} = V_{CC} I_{DC} = V_{CC} I_{CQ} \tag{26.1}$$

The power provided by the amplifier can be calculated using

$$P_o(\text{AC}) = \frac{V_C^2(\text{rms})}{R_C} = \frac{V_C^2(\text{peak})}{2R_C} = \frac{V_C^2(\text{p-p})}{8R_C} \tag{26.2}$$

with the amplifier's efficiency being

$$\%\eta = 100 \times \frac{P_o(\text{AC})}{P_i(\text{DC})}\% \tag{26.3}$$

A class-B amplifier draws no power if no input signal is applied. As the input signal increases the amount of power drawn from the voltage supply and that delivered to the load both increase. The input power to a class-B amplifier is

$$P_i(\text{DC}) = V_{CC} I_{DC} = \frac{2 V_{CC} V_C(p)}{\pi R_L} \tag{26.4}$$

The amplifier efficiency is calculated using Eq. (26.3).

$$P_o(\text{AC}) = \frac{V_L^2(\text{rms})}{R_L} = \frac{V_L^2(p)}{2R_L} = \frac{V_L^2(\text{p-p})}{8R_L} \tag{26.5}$$

PROCEDURE

Part 1. Class-A Amplifier: DC Bias

a. Calculate DC bias values for the circuit of Fig. 26.1.

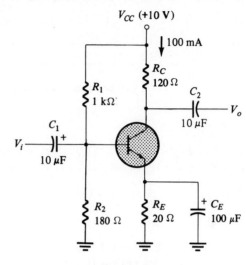

Figure 26-1

(calculated) V_B = _____
(calculated) V_E = _____
(calculated) $I_E = I_C$ = _____
(calculated) V_C = _____

b. Construct the circuit of Fig. 26.1. If desired, measure and record actual resistor values in the space provided in Fig. 26.1. Adjust the supply voltage to V_{CC} = 10 V and measure and record DC bias voltages:

(measured) V_B = _____
(measured) V_E = _____
(measured) V_C = _____

Determine the measured value of DC bias current:

$I_E = I_C = V_E/R_E$ = _____

Part 2. Class-A Amplifier: AC Operation

a. Using DC bias values calculated in Part 1 and equations given in review section, calculate power and efficiency values for the largest signal swing in the class-A amplifier of Fig. 26.1.

(calculated) P_i = _____

Using largest signal swing around DC bias set in Part 1:

(calculated) V_o = _____
(calculated) P_o = _____
(calculated) % η = _____

b. Using the oscilloscope adjust the input signal (f = 10 kHz) to obtain the largest undistorted output signal. Measure and record these input and output voltages.

(measured) V_i = _____
(measured) V_o = _____

c. Using measured values calculate the power and efficiency for the class-A amplifier of Fig. 26.1.

P_i = _____
P_o = _____
% η = _____

Compare the measured and calculated values of power and efficiency obtained in steps **b** and **c**.

Procedure

d. Reduce the input signal to one-half the level of step **b**. Measure and record input and output voltages.

(measured) V_i = _____
(measured) V_o = _____

e. Calculate the input power, output power and efficiency using half the input voltage used in step **a**.

(calculated) P_i = _____
(calculated) P_o = _____
(calculated) % η = _____

f. Using measured values calculate the power and efficiency for the class-A amplifier of Fig. 26.1.

P_i = _____
P_o = _____
% η = _____

Compare the measured and calculated values of power and efficiency obtained in steps **e** and **f**.

Part 3. Class-B Amplifier Operation

a. Calculate the power ratings for a class-B amplifier, as in Fig. 26.2 for V_o = 1 V, peak and V_o = 2 V, peak.

(calculated) P_i = _____
(calculated) P_o = _____
(calculated) % η = _____

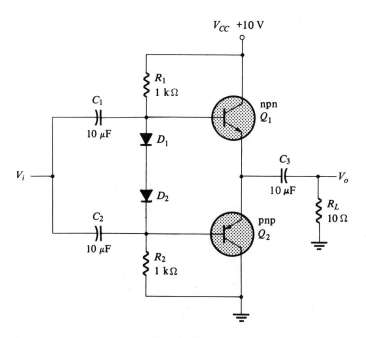

Figure 26-2

For $V_o = 2$ V, peak:

(calculated) P_i = _____
(calculated) P_o = _____
(calculated) % η = _____

b. Construct the circuit of Fig. 26.2. Adjust $V_{CC} = 10$ V. If desired, measure and record actual resistor values in space provided in Fig. 26.2. Adjust input until $V_o = 1$ V, peak. Measure and record AC voltages.

(measured) V_i = _____
(measured) V_o = _____

Procedure

Using measured values calculate input and output power, and circuit efficiency.

$P_i =$ _____
$P_o =$ _____
$\% \eta =$ _____

Compare values calculated in step **a** with those measured in step **b**.

c. Adjust input until $V_o = 2$ V, peak. Measure and record AC voltages.

(measured) $V_i =$ _____
(measured) $V_o =$ _____

Measure average (DC) supply current from V_{CC}.

(measured) $I_{DC} =$ _____

Using measured values, calculate input and output power, and circuit efficiency:

$P_i =$ _____
$P_o =$ _____
$\% \eta =$ _____

Compare values calculated in step **a** with those measured in step **c**.

Name _____
Date _____
Instructor _____

Differential Amplifier Circuits

OBJECTIVE

To measure DC and AC voltages in differential amplifier circuits.

EQUIPMENT REQUIRED

Instruments

Oscilloscope
DMM
Function Generator
DC supply

Components

Resistors

(1) 4.3-kΩ
(4) 10-kΩ
(2) 20-kΩ

Transistors

(3) 2N3823, or equivalent

EQUIPMENT ISSUED

Item	Laboratory serial no.
DC Power Supply	
Function generator	
Oscilloscope	
DMM	

RÉSUMÉ OF THEORY

BJT Differential Amplifier

A differential amplifier is a circuit with plus (+) or minus(−) inputs. In typical operation, inputs that are opposite in-phase are amplified greatly, while inputs that are in-phase are canceled at the output. Figure 27.1 is the circuit of a simple BJT differential amplifier with plus (V_i^+) input and minus (V_i^-) input, and opposite outputs, V_{o1} and V_{o2}. Typically no capacitor is needed, the input signals being DC coupled, and the positive (V_{CC}) and negative (V_{EE}) supplies providing DC bias. Using the value of r_e assumed in this experiment to be the same for both transistors, the differential voltage gain is of magnitude

$$A_v = \frac{R_C}{2r_e} \tag{27.1}$$

The gain for signals which are common at both inputs is of magnitude

$$A_v = \frac{R_C}{2R_E} \tag{27.1}$$

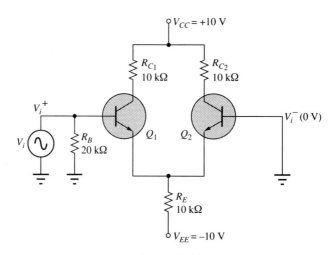

Figure 27-1

FET Differential Amplifier

For an FET differential amplifier the magnitude of the differential voltage gain can be calculated as

$$A_v = \frac{g_m R_D}{2} \tag{27.3}$$

PROCEDURE

Part 1. DC Bias of BJT Differential Amplifier

a. For the circuit of Fig. 27.1 calculate DC bias voltages and currents for one transistor.

(calculated) V_B = _____
(calculated) V_E = _____
(calculated) V_C = _____
(calculated) I_E = _____
(calculated) r_e = _____

b. Construct the circuit of Fig. 27.1. (Record measured value for all resistors in Fig. 27.1.) Set both supplies, V_{CC} = 10 V and V_{EE} = 10 V. Measure and record DC bias voltages for each transistor.

	Q_1	Q_2
(measured) V_B =	_____	V_B = _____
(measured) V_E =	_____	V_E = _____
(measured) V_C =	_____	V_C = _____

Using measured values determine

I_E = _____ I_E = _____
r_e = _____ r_e = _____

Compare values for each transistor to determine if they are well matched. Compare the values calculated in step **a** with those measured in step **b**.

Part 2. AC Operation of BJT Differential Amplifier

a. Using Eqs. (27.1) and (27.2) calculate the differential and common-mode gain of the circuit in Fig. 27.1.

(calculated) $A_{v,d}$ = _____
(calculated) $A_{v,c}$ = _____

b. Apply input of V_i = 20 mV, rms at frequency f = 10 kHz to the plus (+) input and 0 V to the minus (−) input in the circuit of Fig. 27.1. Using a DMM measure, record output voltages.

(measured) V_{o1} = _____
(measured) V_{o2} = _____

Calculate an average value of $V_{o,d}$

$$V_{o,d} = \frac{V_{o,1} + V_{o,2}}{2}$$

$V_{o,d}$ = _____

Calculate differential voltage gain

$$A_{v,d} = \frac{V_{o,d}}{V_i}$$

(measured) $A_{v,d}$ = _____

Procedure

c. Apply common inputs of $V_i = 1$ V, peak to both input terminals in the circuit of Fig. 27.1. Measure and record the output from one side of the circuit

(measured) $V_{v,c}$ = _____

Calculate the common voltage gain

$$A_{v,c} = \frac{V_{o,c}}{V_i}$$

(measured) $A_{v,c}$ = _____

Compare the voltage gains calculated in step **a** with those measured in steps **b** and **c**.

Part 3. DC Bias of BJT Differential Amplifier With Current Source

a. Calculate DC bias voltages and currents for the amplifier of Fig. 27.2.

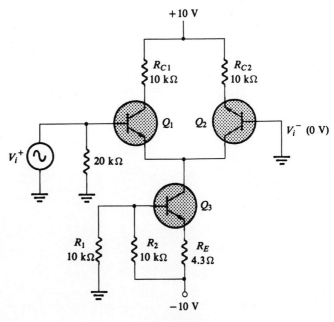

Figure 27-2

For either Q_1 or Q_2:

(calculated) V_B = _____
(calculated) V_E = _____
(calculated) V_C = _____
(calculated) I_E = _____
(calculated) r_e = _____

For Q_3:

(calculated) V_B = _____
(calculated) V_E = _____
(calculated) V_C = _____
(calculated) I_E = _____
(calculated) r_e = _____

b. With DC power turned off, construct the circuit of Fig. 27.2 (or just modify the circuit of Part 2). (Record measured resistor values in Fig. 27.2.) Restore DC power (10 V and −10 V) and measure DC bias voltages.

For both transistor Q_1 and Q_2:

	Q_1	Q_2
(measured) V_B =	_____	V_B = _____
(measured) V_E =	_____	V_E = _____
(measured) V_C =	_____	V_C = _____

Using measured values determine

I_E = _____ I_E = _____
r_e = _____ r_e = _____

Compare values for both transistors to determine if transistors are well matched.

For transistor Q_3:

(measured) $V_B = $ _____
(measured) $V_E = $ _____
(measured) $V_C = $ _____

Using measured values determine

$I_E = $ _____
$r_e = $ _____

Compare the values calculated in step **a** with those measured in step **b**.

Part 4. AC Operation of Differential Amplifier With Transistor Current Source

a. Using Eq. (27.1) calculate

(calculated) $A_{v,d} = $ _____

b. Apply input of $V_i^+ = 10$ mV, rms at frequency $f = 10$ kHz. Measure and record AC voltages.

(measured) $V_{o,d} = $ _____

$$A_{v,d} = \frac{V_{o,d}}{V_i}$$

(measured) $A_{v,d}$ = _____

c. Apply common input of $V_i = 1$ V, rms at frequency $f = 10$ kHz to both input terminals in the circuit of Fig. 27.2. Measure and record the output from one side of the circuit

(measured) $V_{o,c}$ = _____

Calculate the common voltage gain

$$A_{v,c} = \frac{V_{o,c}}{V_i}$$

(measured) $A_{v,c}$ = _____

d. Using the AC coupled input of the oscilloscope, measure and record the waveforms at each output and at the common-emitter point of the circuit. Record the waveforms in Fig. 27.3 showing proper phase relations.

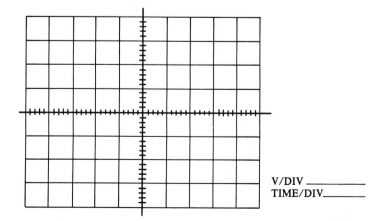

V/DIV _____
TIME/DIV _____

Figure 27-3

Part 5. JFET Differential Amplifier

a. Obtain the values of I_{DSS} and V_P for each of the transistors in the circuit of Fig. 27.4 using previous procedures in Lab. Experiment 12 or 13. Record the values obtained.

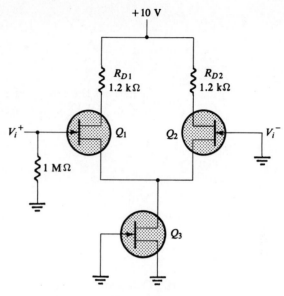

Figure 27-4

For Q_1:

I_{DSS} = _____
V_P = _____

For Q_2:

I_{DSS} = _____
V_P = _____

For Q_3:

I_{DSS} = _____
V_P = _____

b. Calculate the DC bias voltages and currents for the circuit of Fig. 27.4 using the values obtained in step **a.**

(calculated) $V_{D,1}$ = _____
(calculated) $V_{D,2}$ = _____
(calculated) $V_{S,1}$ = _____

c. Construct the circuit of Fig. 27.4. [Record measured resistor values in Fig. 27.4.] Measure and record the DC voltages.

(measured) $V_{G,1}$ = _____
(measured) $V_{D,1}$ = _____
(measured) $V_{D,2}$ = _____
(measured) $V_{D,3}$ = _____

d. Calculate the value of the circuit differential voltage gain.

(calculated) $A_{v,d}$ = _____

e. Using AC coupling, apply an input, V_i^+ = 50 mV, rms at frequency f = 10 kHz. Using a DMM measure, record the output voltages.

(measured) $V_{o,1}$ = _____
(measured) $V_{o,2}$ = _____

Determine AC differential voltage gain (using $V_{o,1}$ and then $V_{o,2}$):

$A_{v1,d}$ = _____
$A_{v2,d}$ = _____

Procedure

Compare the values of differential voltage gain measured in step **e** with that calculated in step **d**.

f. For input of $V_i^+ = 50$ mV, peak observe waveforms at all three transistors drain terminals and record in Fig. 27.5 showing proper phase relations.

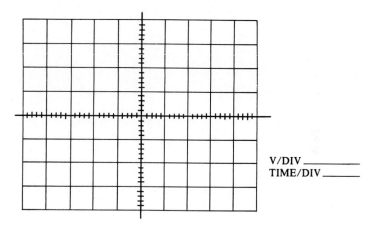

V/DIV _____
TIME/DIV _____

Figure 27-5

$A_{v,c} = $ _____

Part 6. Computer Analysis

a. Use the following PSpice program to analyze the BJT differential amplifier of Fig. 27.1.

```
BJT Differential Amplifier Circuit
VIP    1   0    SIN (0  10mV  1KHz  0  0)
VIM    5   0    SIN (0  0MV   1kHz  0  0)
VCCP   6   0    +10V
VCCM   7   0    -10V
RC1    6   3    10K
RC2    6   4    10K
RE     2   7    10K
Q1     3   1    2   QN
Q2     4   5    2   QN
.MODEL QN NPN (BF = 100)
.OP
.TRAN 0.1ms 2ms
.PLOT TRAN V(3) V(4)
.OPTIONS NOPAGE
.END
```

b. Modify the above listing to analyze the circuit of Fig. 27.2 and run to obtain DC bias and AC voltages.

c. Write a PSpice program to analyze the JFET amplifier of Fig. 27.4 and run to obtain DC bias values and show AC waveforms in circuit.

Name _____
Date _____
Instructor _____

EXPERIMENT 28
Linear Op-Amp Circuits

OBJECTIVE

To measure DC and AC voltages in linear op-amp circuits.

EQUIPMENT REQUIRED

Instruments

Oscilloscope
DMM
Function Generator
DC Supply

Components

Resistors

(1) 20-kΩ
(3) 100-kΩ

ICs

(1) 741 Op-amp

EQUIPMENT ISSUED

Item	Laboratory serial no.
DC Power Supply	
Function generator	
Oscilloscope	
DMM	

RÉSUMÉ OF THEORY

The op-amp is a very high gain amplifier with inverting and noninverting inputs. It can be used to provide a much smaller but exact gain set by external resistors or to sum more than one input, each input having a desired voltage gain.

As an inverting amplifier the resistors are connected to the inverting input as shown in Fig. 28.1 with output voltage

$$V_o = \frac{R_o}{R_i} V_i \tag{28.1}$$

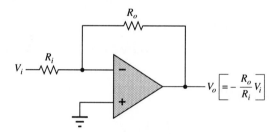

Figure 28-1

A noninverting amplifier is provided by the circuit of Fig. 28.2 with output voltage given by

$$V_o = \left(1 + \frac{R_o}{R_i}\right) V_i \tag{28.2}$$

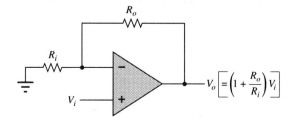

Figure 28-2

Connecting the output back to the inverting input as in Fig. 28.3 provides a gain of exactly unity:

$$V_o = V_i \tag{28.3}$$

Procedure

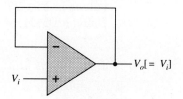

Figure 28-3

More than one input can be connected through separate resistors as shown in Fig. 28.4, with the output voltage then

$$V_o = -\left(\frac{R_o}{R_1}V_1 + \frac{R_o}{R_2}V_2\right) \qquad (28.4)$$

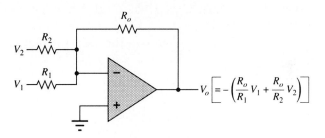

Figure 28-4

PROCEDURE

Part 1. Inverting Amplifier

a. Calculate the voltage gain for the amplifier circuit of Fig. 28.5.

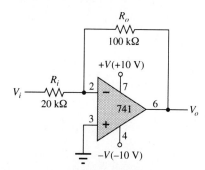

Figure 28-5

(calculated) $V_o/V_i =$ _____

b. Construct the circuit of Fig. 28.5. (Measure and record resistor values in Fig. 28. 5.) Apply an input of $V_i = 1$ V, rms ($f = 10$ kHz). Using a DMM measure and record output voltage.

(measured) V_o = _____

Calculated voltage gain using measured values:

A_v = _____

Compare the gain calculated in step **a** with that measured in step **b**.

c. Replace R_1 with a 100-kΩ resistor. Calculate V_o/V_i.

(calculated) V_o/V_i = _____

For input of $V_i = 1$ V, rms measure and record V_o.

(measured) V_o = _____

Calculate A_v.

A_v = _____

Compare calculated and measured values of voltage gain.

Procedure

d. Using the oscilloscope, observe and sketch input and output waveforms in Fig. 28.6.

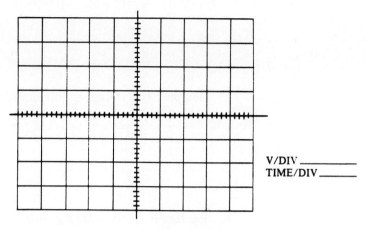

V/DIV _____
TIME/DIV _____

Figure 28-6

Part 2. Noninverting Amplifier

a. Calculate the voltage gain of the noninverting amplifier in Fig. 28.7.

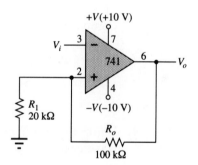

Figure 28-7

(calculated) $A_v = $ _____

b. Construct the circuit of Fig. 28.7. Apply an input of $V_i = 1$ V, rms ($f = 10$ kHz). Using a DMM, measure and record output voltage.

(measured) $V_o = $ _____

Calculate the voltage gain of the circuit using measured voltages.

$V_o/V_i = $ _____

Compare the voltage gain calculated in step **a** with that measured in step **b**.

c. Replace R_1 with a 100 kΩ resistor and repeat steps **a** and **b**.

(calculated) $A_v = $ _____
(measured) $V_o = $ _____
$V_o/V_i = $ _____

Compare the calculated voltage gain with that measured.

d. Using the oscilloscope, observe and sketch the input and output waveforms in Fig. 28.8.

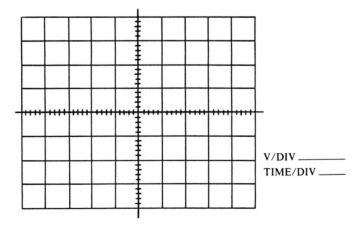

Figure 28-8

Procedure

Part 3. Unity-Gain Follower

a. Construct the circuit of Fig. 28.9. Apply an input signal of $V_i = 2$ V, rms ($f = 10$ kHz). Using a DMM measure and record the input and output voltages.

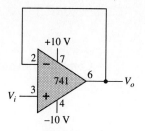

Figure 28-9

(measured) $V_i =$ _____
(measured) $V_o =$ _____

Compare the circuit voltage gain, V_o/V_i with the theoretical unity gain.

Part 4. Summing Amplifier

a. Calculate the output voltage for the circuit of Fig. 28.10 with inputs of $V_1 = V_2 = 1$ V, rms.

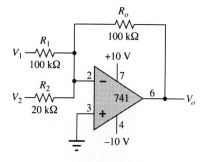

Figure 28-10

(calculated) $V_o =$ _____

b. Construct the circuit of Fig. 28.10. Apply inputs of $V_1 = V_2 = 1$ V, rms ($f = 10$ kHz). Measure and record output voltage.

(measured) $V_o =$ _____

Compare output voltage calculated in step **a** and that measured in step **b**.

c. Change R_2 to 100 kΩ. Repeat steps **a** and **b**.

(calculated) $V_o =$ _____
(measured) $V_o =$ _____

Compare calculated output voltage with that measured.

Name _____
Date _____
Instructor _____

Active Filter Circuits

OBJECTIVE

To measure AC voltages as a function of frequency in various type active filter circuits.

EQUIPMENT REQUIRED

Instruments

Oscilloscope
DMM
Function Generator
DC supply

Components

Resistors

(5) 10-kΩ
(1) 100-kΩ

Capacitors

(2) 0.001-μF

Transistors and ICs

(1) 301 IC, or equivalent

311

EQUIPMENT ISSUED

Item	Laboratory serial no.
DC Power Supply	
Function generator	
Oscilloscope	
DMM	

RÉSUMÉ OF THEORY

Op-amps can be used to build active filter circuits for use as low-pass, high-pass, or brand-pass filter operation. Filter operation provides the output of the filter drop off as a function of frequency to 0.707 of the starting value at the cutoff frequency. This is a drop of 3 dB. The rate of amplitude decrease is at 6-dB per octave (half or twice frequency), which is the same as 20-dB per decade (ten-fold larger or smaller frequency).

Low-Pass Filter

A low-pass active filter passes frequencies below the filter cutoff frequency. The circuit of Fig. 29.1 shows the connection of an op-amp unit as a low-pass filter, the low-cutoff frequency determined by

$$f_L = \frac{1}{2\pi R_1 C_1} \tag{29.1}$$

the output then dropping off at 6 dB/octave or 20 dB/ decade.

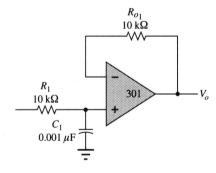

Figure 29-1

High-Pass Filter

A high-pass filter, as that of Fig. 29.2, maintains the output amplitude at frequencies above a high cutoff frequency determined by

$$f_H = \frac{1}{2\pi R_2 C_2} \tag{29.2}$$

Procedure

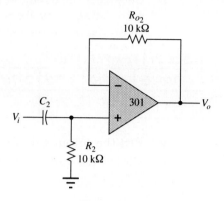

Figure 29-2

Band-Pass Filter

A band-pass filter circuit, as shown in Fig. 29.3, passes the input signal only for frequencies within a band of frequencies. The circuit shown is basically low-pass and high-pass active filters in series. The band-pass low and high cutoff frequencies are then calculated using Eqs. (29.1) and (29.2).

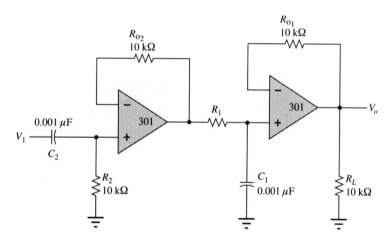

Figure 29-3

PROCEDURE

Part 1. Low-Pass Active Filter

    a. For the circuit of Fig. 29.1 calculate the low cutoff frequency using Eq. (29.1).

(calculated) $f_L =$ _____

b. Construct the circuit of Fig. 29.1. Apply input of 1 V, rms. Vary the signal frequency from 100 Hz to 50 kHz while measuring and recording the output voltage in Table 29.1.

TABLE 29.1 Low-Pass Filter

f	100-Hz	500-Hz	1-kHz	2-kHz	5-kHz	10-kHz	15-kHz	20-kHz	30-MHz
V_o									

c. Plot the output gain-frequency response curve in Fig. 29.4.

d. Obtain the value of low-cutoff frequency from the data plotted in Fig. 29.4.

(measured) $f_L = $ _____

Compare the low-cutoff frequency calculated in step **a** with that obtained in step **d**.

Part 2. High-Pass Active Filter

a. Using Eq. (29.2), calculate the high-cutoff frequency for the circuit of Fig. 29.2.

b. Construct the circuit of Fig. 29.2. Apply an input of 1 V, rms. Vary the signal frequency from 1 kHz to 300 kHz and record the resulting output voltage in Table 29.2.

TABLE 29.2 High-Pass Filter

f	1-kHz	2-kHz	5-kHz	10-kHz	20-kHz	30-kHz	50-kHz	100-kHz	300-kHz
V_o									

c. Plot the data obtained in Table 29.2 in Fig. 29.5.

d. Using the plot in Fig. 29.5 obtain the high-cutoff frequency.

(measured) $f_H = $ _____

Compare the high cutoff frequency calculated in step **a** with that measured in step **d**.

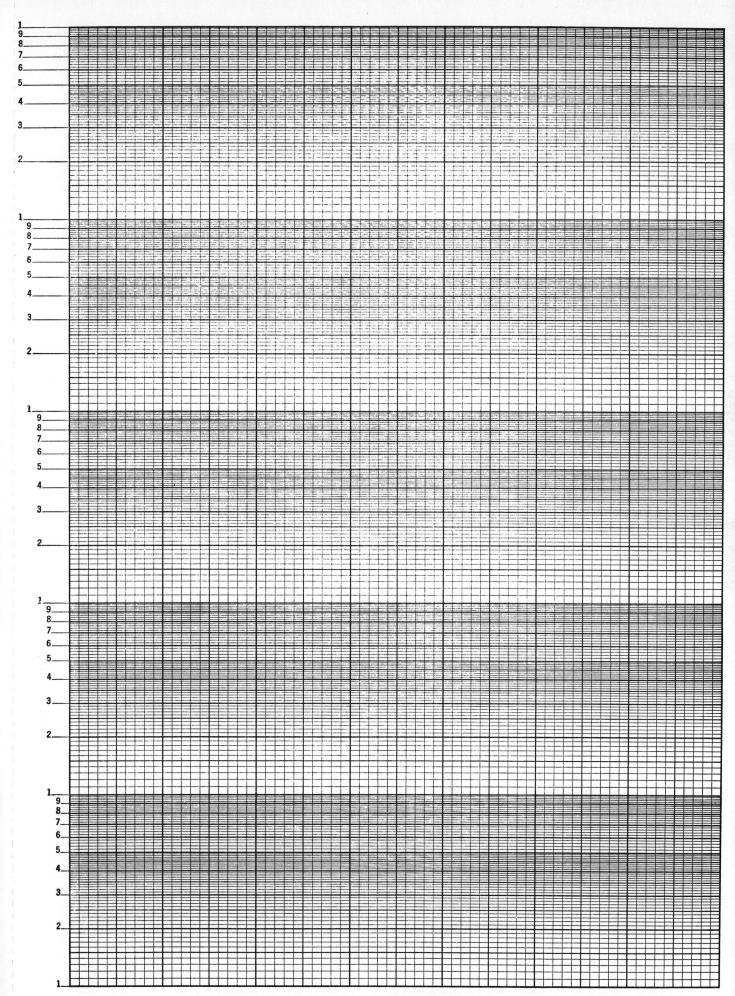
Figure 29-4

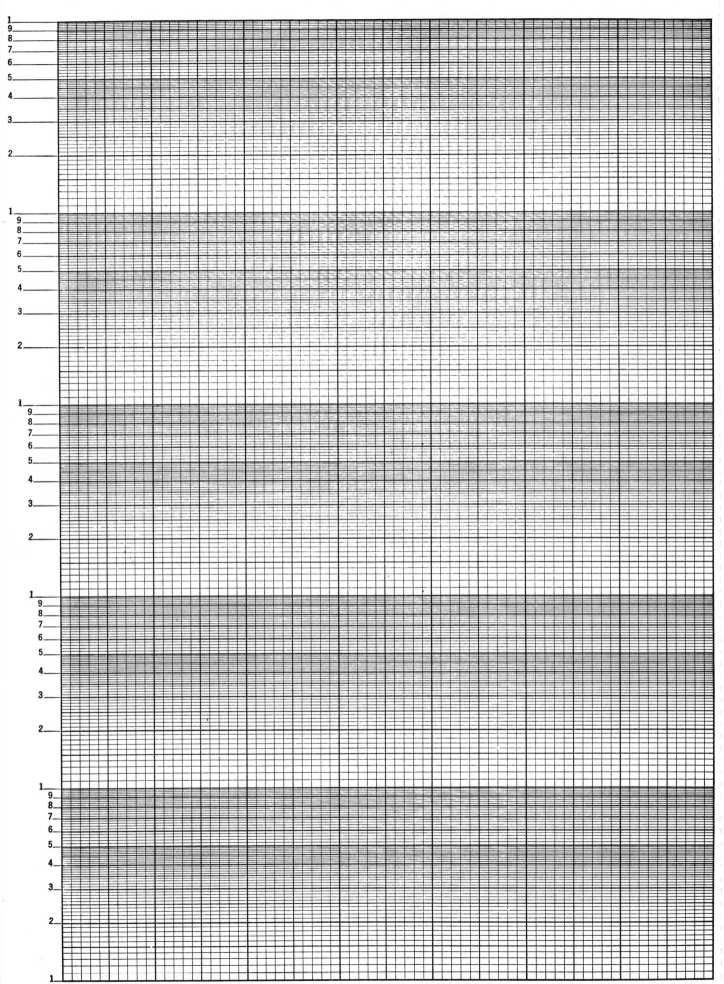

Figure 29-5

Procedure

Part 3. Band-Pass Active Filter

a. Calculate the band-pass frequencies using Eqs. (29.1) and (29.2).

b. Construct the circuit of Fig. 29.3.

c. Apply an input signal of 1 V, rms. Vary the signal frequency from 100 Hz to 300 kHz and record output voltage in Table 29.3.

TABLE 29.3 Band-Pass Filter

f	100-Hz	500-Hz	1-kHz	2-kHz	5-kHz	10-kHz	15-kHz	20-kHz	30-MHz
V_o									

f	50-kHz	100-kHz	200-kHz	300-kHz
V_o				

d. Plot the data in Fig. 29.6. Using the plot determine the lower and higher cutoff frequencies for the band-pass filter.

Compare the frequencies calculated in step **a** with those measured in step **d**.

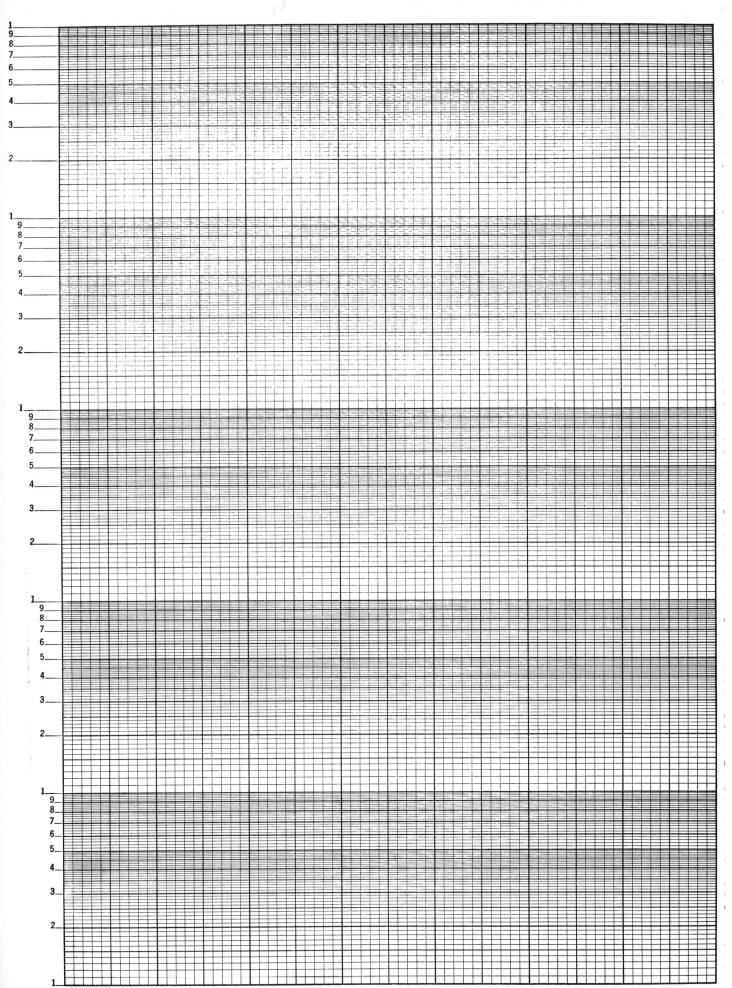

Figure 29-6

EXPERIMENT 30

Comparator Circuit Operation

OBJECTIVE

To measure DC and AC operation using comparator IC circuits.

EQUIPMENT REQUIRED

Instruments

Oscilloscope
DMM
Function Generator
DC Supply

Components

Resistors

(1) 1-kΩ
(1) 3.3-kΩ
(3) 10-kΩ
(1) 20-kΩ
(3) 100-kΩ
(1) 50-kΩ potentiometer

Capacitors

(2) 15-μF
(1) 100-μF

Transistors and ICs

(1) 2N3904
(1) 741 Op-amp IC (or equivalent)
(1) 339 Comparator IC (or equivalent)
(1) LED (20 mA)

EQUIPMENT ISSUED

Item	Laboratory serial no.
DC Power Supply	
Function generator	
Oscilloscope	
DMM	

RÉSUMÉ OF THEORY

A comparator circuit is essentially a very high gain op-amp having a plus (+) and a minus (−) input. The output of the comparator is a logic level that provides an indication of when the plus input voltage is greater than the minus input or when the plus input is less than the minus input. Although an op-amp can be used for this purpose, special comparator ICs are available which are better suited for this operation.

Figure 30.1 shows a 741 op-amp used as a level detector. The reference level voltage V_{ref} is set at +5 V. The indicator LED goes *on* whenever the input V_i goes *below* V_{ref} and goes *off* whenever V_i goes *above* V_{ref}. Figure 30.2 shows a similar operation using a 339 comparator IC.

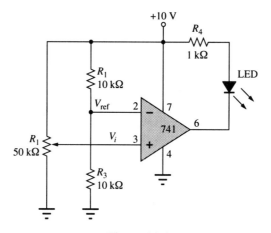

Figure 30-1

Figure 30.3 shows two comparator stages connected as a window detector—a circuit which provides indication of whenever the input voltage is *within* a specified range of voltage.

Procedure

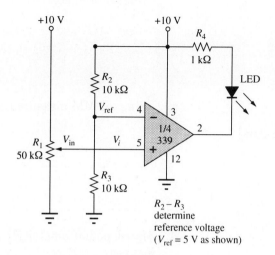

Figure 30-2

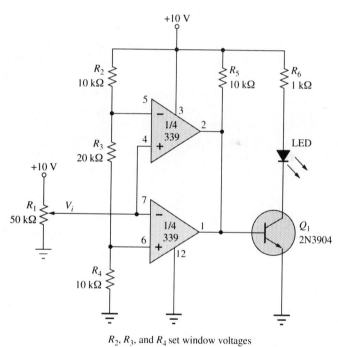

R_2, R_3, and R_4 set window voltages

Figure 30-3

PROCEDURE

a. For the circuit of Fig. 30.1 calculate V_{ref}.

(calculated) $V_{ref} = $ _____ ($R_3 = 10$ kΩ)

b. Construct the circuit of Fig. 30.1. (Measure and record resistor values in Fig. 30.1.)

c. Using a DMM measure the reference voltage, V_{ref}.

(measured) V_{ref} = _____

d. Adjust potentiometer R_1 so that the LED just goes *on*[*], and then just goes *off*. Record the voltage V_i for each condition.

(measured) V_i (LED goes *on*) = _____
(measured) V_i (LED goes *off*) = _____

e. Replace R_3 with a 20 kΩ resistor and repeat steps **b** and **c**.

(measured) V_{ref} = _____
(measured) V_i (LED goes *on*) = _____
(measured) V_i (LED goes *off*) = _____

Compare the values of V_{ref} calculated in step **a** with those measured in steps **c** and **e**.

---

[*]The potentiometer could be replaced by a rectified triangular wave at 100 Hz to provide the V_i input. The V_i signal and comparator V_o can then be observed simultaneously on the oscilloscope.

Procedure

Part 2. Comparator IC Used as Level Detector

a. For the circuit of Fig. 30.2 calculate V_{ref}.

(calculated) V_{ref} = _____ (R_3 = 10 kΩ)

Repeat calculation for R_3 = 20 kΩ.

(calculated) V_{ref} = _____ (R_3 = 20 kΩ)

b. Construct the circuit of Fig. 30.2. (Measure and record resistor values in Fig. 30.2.)

c. Using a DMM measure the reference voltage.

(measured) V_{ref} = _____ (R_3 = 10 kΩ)

d. Adjust potentiometer R_1 so that the LED just goes *on** and also just goes *off*. Measure the input voltage for each condition.

(measured) V_i = _____ (LED goes *on*)

(measured) V_i = _____ (LED goes *off*)

e. Replace R_1 with a 20 kΩ resistor. Repeat steps **c** and **d**.

(measured) V_{ref} = _____ (R_3 = 20 kΩ)

(measured) V_i = _____ (LED goes *on*)

(measured) V_i = _____ (LED goes *off*)

---

*see footnote on page 322.

f. Interchange connections at pins 4 and 5 so that V_i goes to the *minus* input and V_{ref} goes to the *plus* input. Repeat step **d.**

(measured) V_i = _____ (LED goes *on*)

(measured) V_i = _____ (LED goes *off*)

Compare the calculated and measured voltages in steps **c** through **f** with that calculated in step **a.**

Part 3. Window Comparator

a. For the circuit of Fig. 30.3 calculate V^+ (pin 5) and V^- (pin 6).

(calculated) V^+ (pin 5) = _____
(calculated) V^- (pin 6) = _____

b. Construct the circuit of Fig. 30.3. (Measure and record resistor values in Fig. 30.3.)

c. Using a DMM measure the voltages at pins 1, 5, and 6.

(measured) V_i (pin 1) = _____
(measured) V^+ (pin 5) = _____
(measured) V^- (pin 6) = _____

d. Adjust V_i from 0 V to +10 V*. Measure the voltage levels at which the LED goes *on* and then goes *off*.

(measured) V_i = _____ (LED goes *on*)

(measured) V_i = _____ (LED goes *off*)

e. Adjust V_i from +10 V to 0 V. Measure the voltage levels at which the LED goes *on* and then goes *off*.

(measured) V_i = _____ (LED goes *on*)

(measured) V_i = _____ (LED goes *off*)

f. Interchange resistors R_3 and R_4 and repeat step **d**.

(measured) V_i = _____ (LED goes *on*)

(measured) V_i = _____ (LED goes *off*)

Compare the calculated values in step **a** with those measured in step **c**.

---

*see footnote on page 322.

Name _____
Date _____
Instructor _____

EXPERIMENT 31

Oscillator Circuits

OBJECTIVE

To measure waveforms in various oscillator circuits.

EQUIPMENT REQUIRED

Instruments

Oscilloscope
DMM
Function Generator
DC Supply

Components

Resistors

(3) 10-kΩ
(1) 51-kΩ
(1) 100-kΩ
(1) 220-kΩ
(1) 500-kΩ pot

Capacitors

(3) 0.001-μF
(3) 0.01-μF
(1) 15-μF

ICs

(1) 7414 Schmitt Trigger IC
(1) 741(or equivalent) op-amp
(1) 555 Timer IC

EQUIPMENT ISSUED

Item	Laboratory serial no.
DC Power Supply	
Function generator	
Oscilloscope	
DMM	

RÉSUMÉ OF THEORY

Oscillator circuits can be built using op-amps with feedback to phase-shift the output signal by 180°. Phase-shift: In a phase-shift oscillator, as shown in Fig. 31.1 three sections of resistor-capacitor are used. The resulting oscillator frequency can be calculated using

$$f = \frac{1}{2\pi\sqrt{6}\,RC} \tag{31.1}$$

For the op-amp to cause the circuit to oscillate requires that the op-amp gain be of magnitude 29.

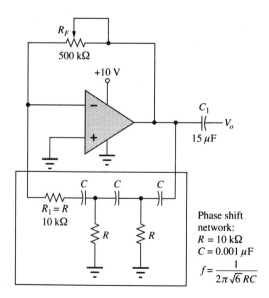

Figure 31-1

Wien-Bridge

A bridge network can be used to provide the 180° phase shift as shown in Fig. 31.2. The circuit's resulting frequency can be calculated from

$$f = \frac{1}{2\pi\sqrt{R_1 C_1 R_2 C_2}} \tag{31.2}$$

Résumé of Theory

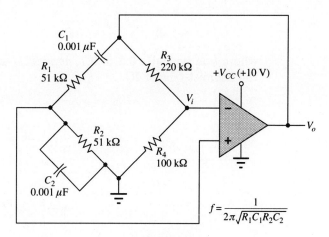

Figure 31-2

If $R_1 = R_2 = R$, and $C_1 = C_1 = C$, then

$$f = \frac{1}{2\pi RC} \tag{31.3}$$

Square-wave Oscillator

A 555 Timer IC is a versatile linear digital IC which can be wired for operation as an oscillator, as shown in Fig. 31.3. The output resulting from this circuit is a pulse clock waveform of frequency

$$f = \frac{1.44}{(R_A + 2R_B)C} \tag{31.4}$$

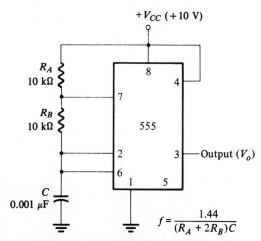

Figure 31-3

Schmitt-Trigger Oscillator

A single Schmitt-trigger IC, resistor, and capacitor can be used to build a pulse-type oscillator circuit, as shown in Fig. 31.4. The oscillator frequency is generally calculated using

$$f = \frac{k}{RC} \tag{31.5}$$

where k is typically 0.3 to 0.7, depending on the internal triggering levels of the Schmitt-trigger IC.

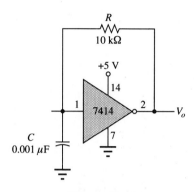

Figure 31-4

PROCEDURE

Part 1. Phase-Shift Oscillator

a. Construct the circuit of Fig. 31.1 with a R_F = 500 kΩ potentiometer, R_1 = 22 kΩ, R = 100 kΩ and C = 0.001 µF. (Measure and record resistor values in Fig. 31.1.)

b. Use the oscilloscope to record the output waveform of the oscillator circuit in Fig. 31.5. Adjust R_F for maximum undistorted output waveform, V_o. Record value of R_F for this undistorted condition.

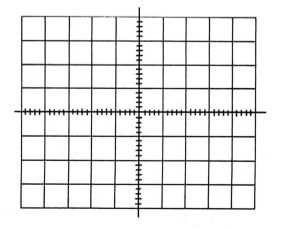

Figure 31-5

Procedure

(measured) R_F = _____

c. Measure and record the time for one cycle of the waveform.

(measured) Period, T = _____

d. Determine the frequency of the waveform.

Frequency, $f = 1/T$ = _____

e. Replace the capacitors with $C = 0.01$ µF and repeat steps **c** through **d**.

(measured) Period, T = _____
Frequency, $f = 1/T$ = _____

f. Calculate the theoretical frequency using Eq. (31.1) for both capacitor values.

(calculated) $f(C = 0.001$ µF$)$ = _____
(calculated) $f(C = 0.01$ µF$)$ = _____

Compare the measured and calculated frequencies for both capacitor values.

Part 2. Wien Bridge Oscillator

a. Construct the circuit of Fig. 31.2. (Measure and record resistor values in Fig. 31.2.)

b. Using the oscilloscope observe and record the output waveform in Fig. 31.6.

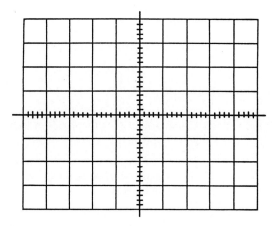

Figure 31-6

c. Measure the time for one cycle.

(measured) $T = $ _____

d. Determine the signal frequency.

$f = 1/T = $ _____

e. Change both capacitors to $C = 0.01$ µF and repeat steps **c** through **d**.

(measured) $T = $ _____
$f = 1/T = $ _____

f. Calculate the theoretical frequency of the oscillator for each capacitor value.

(calculated) $f(C = 0.001~\mu F) = $ _____
(calculated) $f(C = 0.01~\mu F) = $ _____

Compare the calculated frequencies for both capacitor values with those measured.

Part 3. 555 Timer Oscillator

a. Construct the oscillator circuit of Fig. 31.3. (Measure and record resistor values in Fig. 31.3.)

b. Observe and record the output waveforms at pins 3 and 4 in Fig. 31.7.

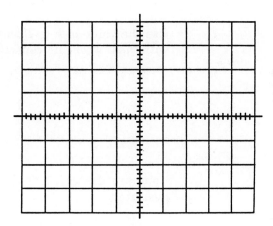

Figure 31-7

c. Measure the period of the output waveform.

(measured) $T = $ _____

d. Determine the signal frequency.

$$f = 1/T = \underline{\hspace{2cm}}$$

e. Replace the capacitor with $C = 0.01$ µF, and repeat steps **c** through **d**.

$$\text{(measured) } T = \underline{\hspace{2cm}}$$
$$f = 1/T = \underline{\hspace{2cm}}$$

f. Calculate the value of coefficient k using the measured values of f, R and C.

$$\text{(measured) } k = \underline{\hspace{2cm}}$$

Use the value of k to calculate the theoretical value of frequency with the $C = 0.01$ µF capacitor.

$$\text{(calculated) } f = \underline{\hspace{2cm}}$$

Compare the calculated value of f for $C = 0.01$ µF with that measured.

Part 4. Schmitt-trigger Oscillator

a. Construct the oscillator circuit of Fig. 31.4. (Measure and record resistor value in Fig. 31.4.)

b. Observe and record the output waveforms at pins 1 and 2 in Fig. 31.8.

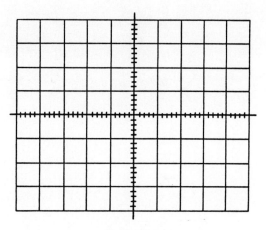

Figure 31-8

c. Measure the period of the output waveform.

(measured) $T =$ _____

d. Determine the signal frequency.

$f = 1/T =$ _____

e. Replace the capacitor with $C = 0.01$ µF, and repeat steps **c** through **d**.

(measured) $T =$ _____
$f = 1/T =$ _____

f. Calculate the theoretical frequency, f, using Eq. (31.5) for each of the capacitor values.

(calculated) $f(C = 0.001$ µF$) =$ _____
(calculated) $f(C = 0.01$ µF$) =$ _____

Compare the calculated value of f for each capacitor with those measured.

Name _____
Date _____
Instructor _____

EXPERIMENT 32

Voltage Regulation-Power Supplies

OBJECTIVE

To measure DC and ripple voltages in series and shunt regulator circuits.

EQUIPMENT REQUIRED

Instruments

Oscilloscope
DMM
Function Generator
DC supply

Components

Resistors

(1) 390 Ω, 2 W
(2) 1-kΩ
(1) 2-kΩ
(1) 20-kΩ
(1) 100-kΩ

Transistors & ICs

(1) *npn* power transistor
(1) op-amp (741 or equivalent)

EQUIPMENT ISSUED

Item	Laboratory serial no.
DC Power Supply	
Function generator	
Oscilloscope	
DMM	

RÉSUMÉ OF THEORY

Voltage regulators attempt to maintain a constant DC output voltage by controlling the series current fed to the load—series voltage regulation, or by controlling the current to the load by shunting some away—shunt regulation.

Series Regulation

The circuit of Fig. 32.1 shows a basic series regulator circuit. The Zener diode provides a reference voltage which sets the output voltage at

$$V_L = V_Z - V_{BE} \tag{32.1}$$

If the output voltage tends to go lower, the series transistor is driven further into conduction providing more current to the load to maintain the output voltage.

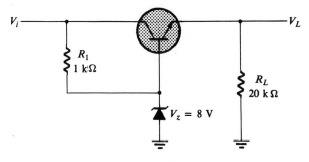

Figure 32-1

Improved Series Regulation

The circuit of Fig. 32.2 shows the addition of an op-amp to provide improved regulation. The output voltage is set by the zener diode and feedback network made of resistors R_1 and R_2. The voltage gain of the op-amp, connected in a positive-feedback configuration is

$$A = \frac{R_1}{R_2} \tag{32.2}$$

with

$$V_L = AV_Z \tag{32.3}$$

If the output voltage tends to get larger, the increased feedback voltage sensed by voltage divider R_1 and R_2 causes a reduced input to the op-amp, less drive current to the series pass transistor, and reduced load current, thereby maintaining the output voltage.

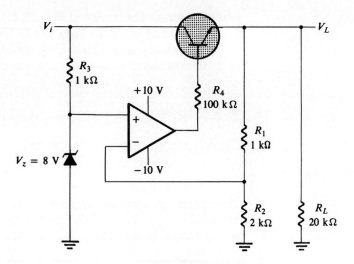

Figure 32-2

Shunt Regulation

The circuit of Fig. 32.3 shows a transistor connected in parallel (shunt) with the output. The transistor conducts to provide greater or less load current, thereby maintaining the output voltage. Again, a sensing network made of resistor voltage divider (using R_1 and R_2), controls the input to the op-amp which then controls the conduction of the shunt transistor. The regulated output voltage can be calculated using

$$V_L = \frac{R_1 + R_2}{R_1} V_Z \qquad (32.4)$$

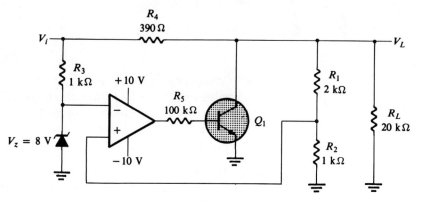

Figure 32-3

PROCEDURE

Part 1. Series Voltage Regulator

a. Calculate the resulting regulated voltage for the circuit of Fig. 32.1.

b. Construct the circuit of Fig. 32.1. (Measure and record the resistor values in Fig. 32.1.) Vary the DC input voltage, V_i, from 10 V to 16 V, measuring and recording the load voltage in Table 32.1. Record the regulated output voltage measured.

(measured) $V_o =$ _____

TABLE 32.1 Series Voltage Regulator

V_i	10-V	11-V	12-V	13-V	14-V	15-V	16-V
V_o							

Compare the regulation voltage obtained in step **b** with that calculated in step **a**.

Part 2. Improved Series Regulator

a. Calculate the regulated output voltage for the circuit of Fig. 32.2.

(calculated) $V_L =$ _____

b. Construct the circuit of Fig. 32.2. (Measure and record the resistor values in Fig. 32.2.) Vary the dc input voltage, V_i, from 10 V to 24 V, in 2-V steps, measuring and recording the load voltage, V_L in Table 32.3. Record the value of regulated load voltage.

(measured) $V_L =$ _____

TABLE 32.2 Series Voltage Regulator

V_i	10-V	12-V	14-V	13-V	16-V	18-V	20-V	22-V	24-V
V_L									

Compare the regulation voltage obtained in step **b** with that calculated in step **a**.

Part 3. Shunt Voltage Regulator

a. Calculate the regulated voltage from the circuit of Fig. 32.3.

(calculated) $V_L =$ _____

b. Construct the circuit of Fig. 32.3. (Measure and record the resistor values in Fig. 32.3.) Apply an input voltage varied from 24 V to 36 V steps. Measure and record the load voltage in Table 32.3. Record the regulated output voltage.

(measured) $V_L =$ _____

TABLE 32.3 Series Voltage Regulator

V_i	24-V	26-V	28-V	30-V	32-V	34-V	36-V
V_o							

Compare the regulation voltage obtained in step **b** with that calculated in step **a**.